Early Hominid Scavenging Opportunities

Implications of Carcass Availability in the Serengeti and Ngorongoro Ecosystems

Robert J. Blumenschine

BAR International Series 283
1986

B.A.R.

5, Centremead, Osney Mead, Oxford OX2 0DQ, England.

B.A.R. -S283, 1986: 'Early Hominid Scavenging Opportunities'

ISBN 9780860543657 paperback
ISBN 9781407344683 e-book
DOI https://doi.org/10.30861/9780860543657
A catalogue record for this book is available from the British Library
This book is available at www.barpublishing.com

To Mindi and Grier

TABLE OF CONTENTS

LIST OF TABLES

Table

Table

LIST OF FIGURES

Figure

PREFACE

This monograph represents the main results of a three year project that resulted in a Ph.D. dissertation at the University of California, Berkeley, in 1985. Alterations to the original work are minor with the exception of the addition of more recent census data in Chapter 2 and an expansion of discussion on the archaeological recognition of scavenging in Chapter 5. Like the original, this work emphasizes the ecology of scavenging opportunities as seen in the Serengeti and Ngorongoro. Numerous other data were collected during the eleven month field stint which bear heavily on the taphonomy of scavenging, and hence the potential significance of scavenging to early hominids. A synthesis of ecological observations such as that presented here is seen as a necessary first step in any analysis of prehistoric subsistence, including potential modes of meat and marrow acquisition by early hominids.

In general, researchers attempting to reconstruct prehistoric lifestyles place too little emphasis on first gaining a detailed understanding of ecological process. I make this statement at the risk of slighting the efforts and achievements of the growing number of paleoanthropologists who are aware of this need. But too often still, behavioral interpretations of prehistoric data proceed without due regard for the ecological basis of the system in question, leaving hominid behavior in a vacuum, devoid of inputs from the greater physical and biotic environment in which they functioned. As Bewes stated over fifty years ago:

> The viewpoint of no single one of the human sciences is broad enough to explain fully why man is as he is and why he behaves as he does. Ecology alone stands apart. It is not a branch of science - it is only a viewpoint. The student of man and his works, whether he calls himself archaeologist, ethnologist, anthropologist...or what you will, would do well to become more of an ecologist and try to adopt the more general viewpoint (1931:11).

Since 1931, Bewes' plea has been heeded in many of the sub-disciplines of anthropology. But in paleoanthropology, where we now have the means to reconstruct paleoenvironments in some detail, and where much work has been done to this end, the essential need to link hominids to the emerging paleoecological picture has not been adequately satisfied. No small portion of the deficiency arises from a poor understanding of the ecology - as observable in modern settings - of the systems under investigation. This is why I devote the initial round of analysis of field data on scavenging, much heavier overall on taphonomy and related bone studies than the current work reflects, to an understanding from an archaeologist's point of view of the many factors that affect scavenging opportunities in contemporary settings. A world of effort in taphonomy cannot hope to illuminte accurately or fully our understanding of early hominid lifestyles unless the basic ecology of the systems under study is first understood. This work makes a small contribution to this end.

As essentially a Ph.D. dissertation, this work benefitted from the' contributions of perhaps more individuals than if written by one well along in their academic career. The brevity of my acknowledgments of their contributions below is not meant to underestimate their importance in the completion of the research. Conversely, any inaccuracies and deficiencies that may remain are, of course, my responsibility.

Two people made immeasurably great contributions to the design and completion of the project. Glynn Isaac was integrally involved in all but the final write-up of the study. Before his tragic death, Glynn enthusiastically encouraged and promoted the development of the field study, and discussed its results with an eagerness and insight which I and all of paleoanthropology surely miss. J. Desmond Clark also made invaluable contributions to all stages of the research project. I am particularly grateful for his many letters to funding agencies and to me in the field, which are responsible in no small way for the success of the study.

The work in Tanzania was conducted under the auspices of the Tanzania Antiquities Division, to whom I am grateful for providing the invaluable logistical and research assistance of Prosper Ndessokia and Charles Saanane. The Serengeti Wildlife Research Centre provided housing and a base for field operations, and I thank Coordinator K.N. Hirji, Acting Director M. Bruno, and the staff at the Centre for approving of, and in so many important ways assisting in the field study. I am grateful to the Tanzania National Scientific Research Council for granting a permit to conduct the research, and the authorities of the Tanzania National Parks system and the Ngorongoro Conservation Area Authority for their cooperation.

In the Serengeti, a number of other researchers taught me so many of the secrets of this finest place on earth. I thank for their advice and assistance Alan and Joan Root, Tony Collins, Craig Packer, Anna Pussey, Chris Magin, Sue Canney, Feroz Khurji and especially Tim Caro. As well, my stay in the Serengeti would not have been so enjoyable and productive without my friends at the Seronera Wildlife

Lodge.

In Nairobi, a number of individuals provided assistance during breaks from the field. I thank Richard Leakey, Harry Merrick and Kamoya Kimeu for their help with the vehicle, and Fiona Marshall, Anne Vincent, Henry Bunn, Ellen Kroll, Diane Gifford-Gonzales and Paul Whitehead for logistical assistance and their interest in the project's progress.

The analysis and write-up of results commenced at Berkeley. I thank F. Clark Howell for his discussion on East African fossil carnivores and Thelma Rowell for sharing her knowledge of the East African bush. Many hours of fruitful discussion were spent with Katie Milton, Tim White, John Parkington, Jeanne Sept, Peter Glumac, Curtis Marean and especially Charlie McNutt, who rehashed with me repeatedly appropriate stategies and units of analysis. I am also thankful to the Quantitative Anthropology Laboratory for providing computing funds and facilities, and the expertise of Gene Hammel, Madeline Anderson and Greer Murphy. Write-up was completed at Rutgers University, where the Department of Anthropology provided word processing facilities.

Insightful comments on drafts of this work were received from J.D. Clark, F.C. Howell, T. Rowell and C. Marean. Leonard Blumenschine, jr. made several series of exhaustive and expert editorial comments. Melinda Blumenschine assisted in the labelling of some of the figures. I am also grateful to George Frame for sending census data of Serengeti and Ngorongoro herbivores and carnivores.

Major funding for the field research was provided by the National Science Foundation (BNS 83-20026) to J.D. Clark as Principle Investigator. I am also thankful to N.S.F. for the use of the Land Rover they funded through a grant to Glynn Isaac. The L.S.B. Leakey Foubdation provided a large initial grant and later gave most generously an additional emergency grant to enable the completion of the field study. The Regents of the University of California provided a Travelling Fellowship during the field period.

CHAPTER 1

CHASED OR CHANCED-UPON?

A major behavioral distinction of modern hunter-gatherers among all primates is the important dietary contribution made by tissues from larger mammals. The transition from a mostly vegetarian diet to one marked by regular and nutritionally substantial intakes of meat is commonly viewed as a dietary innovation integral to the evolution of many unique features of human technological abilities and social systems (e.g. Campbell 1979, K. Hill 1982, Isaac 1971, Isaac and Crader 1981, Thompson 1975, 1976, Washburn and Lancaster 1968). Ample evidence for meat-eating is found throughout the whole span of the archaeological record. Fossil bones bearing butchery marks inflicted by the earliest stone-tool-using hominids indicate the dietary transition to have been made, at least in kind, by two million years ago.

Aside from the evidence for meat-eating, the details of the dietary transition are largely unknown. For instance, the proportional contribution of tissues from larger mammals to the whole diet of prehistoric hominids is still a matter of untestable conjecture. Whether meat-eating among pre-sapient hominids was associated with the complex of social adaptations seen among modern hunter-gatherers is also open to question. The prevailing food-sharing model of Early Pleistocene hominid adaptation, involving a distinct sexual division of subsistence labor and the sharing and delayed consumption of gathered foods at a central place or home base (Isaac 1976, 1978a, 1978b), has been challenged recently by a number of researchers (Binford 1981, Potts 1984, Shipman 1983), including Isaac (1983). On a more fundamental level of inquiry, the method(s) used by hominids to acquire meat is still unknown.

Different modes of meat acquisition by omnivorous hominids imply different frequencies of meat-eating and distinct foraging strategies. Two general modes of meat acquisition, hunting and scavenging, are available to a meat-eater. Scavenging, or foraging for and consuming animals found dead, implies meat-eating to be an opportunistic behavior conducted irregularly and with a minimum of social cooperation. Hunting live quarry, on the other hand, denotes meat to be regularly sought and an integral component of hominid dietary strategies; hunting success would be greatly enhanced by social adjustments including a high level of group cooperation during pursuit and capture, if not also during consumption. If the primary mode of meat acquisition by prehistoric hominids can be determined, paleoanthropologists would have a firmer basis for evaluating the nutritional, ranging and social aspects of hominid dietary strategies.

The relevance of methods of meat acquisition to broader questions in prehistory has long been recognized. The earliest interpretations of animal bones found in association with Plio-Pleistocene hominid fossils in the South African cave deposits assumed a primary role of hunting (Dart 1953). The assumption enhanced the humanness of the "ape-men," an essential aspect of Dart's assertion to an initially unconvinced anthropological community of the hominid status of the australopithecines (cf. Binford 1981). Dart's proposal prompted interpretations of early hominid social life as one centered on hunting (Washburn and Lancaster 1968), and characterized by the modern human traits incorporated in the food sharing model (Isaac 1978a, 1978b, Lancaster 1975) and the less honorable killer instinct-based hunting hypothesis (Ardrey 1976). The primacy of hunting gained broad acceptance in the 1960's and early 1970's as field studies of non-human primates documented the small animal meat-eating of chimpanzees and baboons to be achieved almost exclusively by hunting (Harding 1973, Harding and Strum 1976, Hausfater 1976, Strum 1976, Teleki 1973, 1975, 1981, van Lawick-Goodall 1968; see Butinsky 1982 for a review). Hence, aside from pirating baboon kills (Morris and Goodall 1977), only rare scavenging by chimpanzees been reported recently from Tanzania's Mahale Mountains (Hasegawa et al 1983). Baboons also usually avoid scavengeable food (personal observation, S. Washburn pers. comm., but see Hausfater 1976) and have been shown experimentally to avoid any provisioned carcass with which they had no previous experience (Strum 1983). The virtual absence of scavenging among non-human primates has been interpreted to preclude the necessity of a transitional scavenging phase in human evolution (Goodall and Hamburg 1975, Teleki 1975:179). Rather, the hunting of small or incapacitated prey has been seen as an ancient and sufficient preadaptation to hominid predation of larger animals (DeVore and Washburn 1963, Teleki 1981). The essential humanness of two million year old hominids seemed indisputable well into the 1970's.

Scavenging by early hominids, too, has had its long-time proponents. Bartholomew and Birdsell (1953) and J.D. Clark (1959, 1960) first proposed a role for scavenging among early hominids. Their suggestions were based on the apparent lack of anatomical and technological specializations in hominids for hunting especially large animals and the propensity of modern predators to scavenge:

> "Since few meat-eaters are loathe to scavenge, and the implementation which would allow the australopithecines to kill such large animals is not apparent, we

> suggest that scavenging from kills of larger carnivores may have been systematically carried out (Bartholomew and Birdsell 1953:490)."

Likewise, speaking of the makers of the Oldowan, Clark notes:

> "A good deal [of meat] no doubt, especially [from] large animals, could have been obtained by scavenging, and this is still a well-known practice among some primitive human societies today (1960:311-312)."

Clark continues by citing reports of scavenging among Bantu-speaking peoples of the Mashi River region of northern Zimbabwe:

> "Juveniles are still, or were until recently, sent out in the early mornings...to locate where vultures were flying, thus pinpointing the place where a carnivore had made a kill. The villagers would then recover the meat (1960:312)."

These early suggestions did not deny a role for, or even the primacy of, hunting. As such, scavenging was readily incorporated into some general exposes of human evolution (e.g. Howell 1968, Pfeiffer 1969). The seemingly sound reasoning behind Clark's and Bartholomew and Birdsell's suggestion still demands a role for scavenging (e.g. Bunn 1982, Isaac and Crader 1981), but one possibly only supplemental to preying upon the moderately-sized animals that dominate archaeological bone assemblages.

Very few researchers have assigned a primary or exclusive role to scavenging in the evolution of hominid meat-eating. Louis Leakey (1967, in Teleki 1981) proposed the existence of a scavenging phase preliminary to and bereft of hunting. Szalay (1975) suggests the dentition of the australopithecines to be specifically adapted to scavenging. Here, the thickly enameled, large cheek teeth of Pliocene hominids is seen as an adaptation to bone crushing, and their incisiform canines an adaptation for detaching scraps of meat from bones found mostly defleshed. Jonas (1976, contra Wind 1976) cites the ability of humans to become tonically immobilized to discount a predatory ancestry and support an exclusive scavenging position. Little support has been found for any of these suggestions of exclusive scavenging by early hominids. They have, however, served to emphasize a phenomenon known from studies of most modern carnivores (e.g. Schaller and Lowther 1969) and human hunter-gatherers (e.g. Woodburn 1968 for the Hadza and A. Brooks pers. comm. for the San), namely, that hunting and scavenging are not mutually exclusive behaviors.

With scavenging being elevated to at least as realistic a behavior for early hominids as hunting, a number of researchers have recently proposed criteria for distinguishing the two behaviors archaeologically. The proposals represent the first attempts to go beyond speculation and gross analogy to modern meat-eaters, and to test directly the existence of hunting and scavenging on archaeological data. Most criteria have been based on size, age or body part profiles expected to typify fossil bone assemblages accumulated through hunting or scavenging. The criteria have been applied to a number of fossil-bearing sites in eastern and southern Africa spanning much of Pleistocene prehistory.

Most tests of hunting and scavenging yield equivocable and often disparate results, even when applied to the same set of assemblages. This is particularly true of the various South African cave sites (Vrba 1975, 1980, Read-Martin and Read 1975, Shipman and Phillips 1976), for which considerable doubt remains over a more basic issue: the identity of the bone accumulators (Binford and Bertram 1977, Brain 1981, Maguire et al 1980, Richardson 1980). It is also true, however, of sites certainly created at least in part by hominids, including the early Pleistocene sites from Olduvai (Binford 1981, Bunn 1982, Potts 1983, Shipman 1984), and even the Late Pleistocene site of Klasies River Mouth (Klein 1982, Binford 1984).

Two factors, I feel, are responsible for the inconclusiveness of the tests. One factor is philosophical, the other methodological. The philosophical factor pertains to the variable degrees of humanness implied by indications of the primacy of hunting versus scavenging. The traditional view of early Pleistocene hominids as almost human is being eroded by conclusions that scavenging was as least as important as hunting. The most provocative and consistent advocate of the "hardly human" school, Louis Binford, is, in these terms not surprisingly, the strongest advocate of the scavenging position. Binford has challenged Isaac's food-sharing model of early hominid lifestyles since the late 1970's (Binford 1977), and since the early 1980's has been constructing the soundest arguments and most emphatic conclusions in favor of the primacy of scavenging:

> "The signs seem clear. Earliest man, far from appearing as a mighty hunter of beasts, seems to have been the most marginal of scavengers (1983:59)."

Binford's claims have recently culminated in a detailed analysis of the Middle Stone Age fauna from the site of Klasies River Mouth on the South African Cape (Binford 1984). Binford concludes that even early *Homo sapiens* was to a large degree a scavenger, hunting only small ungulates and scavenging the medium and large size forms also represented at the site.

I am not implying that Binford's provocative conclusion is biased by preconceptions of a very late appearance of the full suite of human traits. Nor do I suggest a bias toward the traditional view by those finding evidence for hunting. I do contend, however, that part of the inconclusiveness of the tests stems from an uncertainty reaching to the core of paleoanthropological inquiry: When did we become human?

The methodological factor is, in my view, more important. The criteria for distinguishing hunting from scavenging have an inadequately detailed, and in some cases, inaccurate basis. The deficiency lies primarily in a poor understanding of the ecology of scavenging, rendering statements of its taphonomy and hence archaeological recognition insensitive to important variations in the nature of the scavengeable food supply. Binford (1984), for example, has contributed the most to an understanding of the general taphonomy of scavenging by identifying some of the carcass parts most usually available to a marginal scavenger. His resultant criteria, however, do not accomodate the full range of feeding opportunities a hominid scavenger might have encountered, including a set of edible parts more typical of those available to a predator on a fresh kill. Many of the criteria proposed by other taphonomists, while being based on general principles of behavioral and population biology (most notably Vrba's and Klein's), are also insentitive to the range of meat-eating opportunities potentially encountered by early hominids. For scavenging to be treated as a testable alternative to hunting, a detailed understanding must be acquired of the characteristics of scavenging opportunity. The nature of feeding opportunities for a scavenger will determine that set of preservable traces indicative of scavenging.

There is a second benefit to be realized from a knowledge of the range of feeding opportunities available to a scavenger. Its ultimate application is to the behavioral and social contexts of hominid meat acquisition. A step in this direction can be provided by a knowledge of the ecological contexts among which scavenging success varies. The frequency, yield and contexts of feeding opportunities for a hominid-like scavenger are essential components of models of individual and group foraging strategies (cf. Kurland and Beckerman 1985).

One brief and preliminary study has attempted to assess the feeding opportunities provided by scavenging. Schaller and Lowther (1969) conducted two experiments to illustrate the availability of scavengeable and easily caught prey in the Serengeti National Park. The one experiment on scavenging was conducted over a period of seven days in an area of prey abundance. Schaller and

Lowther traversed the landscape on foot and in a vehicle in search of scavengeable food to approximate the food yields a prehistoric hominid scavenger might encounter. They discovered the minimal remains (marrow bones and brain) of four lion kills of wildebeest and zebra, and substantial amounts of flesh on a naturally dead buffalo. They also encountered two live but severely incapacitated animals which could have been captured easily. Schaller and Lowther conclude that an exclusively meat-eating hominid group could subsist under similar conditions, but this was enabled only by the discovery of the incapacitated prey. Otherwise:

> "The results show that lion kills offer, at best, an erratic source of food, with long lean periods probably broken by occasional finds providing a considerable amount of meat. Our discovery of the dead buffalo was a matter of luck, something that would happen only infrequently (Schaller and Lowther 1969:328)."

Scavenging seems from these results to provide a very irregular and low-yielding source of meat, especially in comparison to hunting. Although few paleoanthropologists would contend that meat was a primary component of early hominid diets, the results imply that only hunting could account for the consistent association of large quantities of animal bone refuse with stone artifacts at archaeological sites. A fundamental question arising from Schaller and Lowther's experiment that did not receive due emphasis is: Are there sufficiently varied feeding opportunities for a hominid-like scavenger to account in part or whole for the compositions of archaeological bone assemblages, or does the taxonomic, age and body part diversity of the assemblages demand the additional or exclusive practice of hunting in their accumulation?

Schaller and Lowther's scavenging experiment, though brief and largely anecdotal, is informative with regard to this question. The experiment shows that hominids could easily acquire scavengeable food, it identified some of the ecological contexts within which the food could be acquired, and it identified some of the carcass types and general series of body parts available to a scavenger. These results provided, in 1969, the very types of criteria useful for delimiting those components of archaeological bone assemblages a scavenging mode of meat acquisition could account for. It also provided useful insight into an ecological context in which scavenging by hominids could be profitable, i.e. during periods of prey abundance (see Chapter 4). Unfortunately, these results were overshadowed by the apparent low yield of the scavenging experiment.

The need to gain a detailed understanding of the scale and characteristics of scavenging opportunity was perceived by this writer in late 1982 in consultation with Glynn Isaac. A more systematic and comprehensive follow-up study to Schaller and Lowther's experiment was designed to identify, through observations of carcass use and availability in contemporary, natural settings, the composition of the scavengeable food supply and the contexts among which the amount and kind of the supply varies. In so doing, one would be in a position to assess accurately the scavenged or hunted origins of archaeological bone assemblages, and to evaluate the type(s) of foraging strategies most likely to have been practiced.

This work is concerned with the following set of questions:

1) How variable is the availability of scavengeable food?
2) What are the sources of the variation?
3) What types of carcasses and carcass tissues are most regularly available to a scavenger in different ecological contexts?
4) What set of scavenging opportunities existing today might a prehistoric hominid scavenger have encountered and profited from?

The two fields of study relevant to these questions, wildlife biology and taphonomy, alone or together do not currently provide sufficiently detailed answers. Some carnivore studies, for example, document rates of whole carcass consumption and frequencies of carcass abandonment (e.g. Eaton 1974, Kruuk 1972, Lamprecht 1981, Schaller 1972). However, the studies do not (and did not intend to) determine the quantity and anatomical location of edible tissues abandoned by predators to scavengers, two fundamental aspects of scavenging opportunity. Similar studies document the nature and influence of competition for carcasses on the feeding success of various carnivore species (e.g. Eaton 1979, Houston 1974, 1979, King 1975, 1976, Kruuk 1967, 1972, Lamprecht 1981, Schaller 1972), but data on the effect of competition from the viewpoint of the carcass itself are wanting. For example, there exists very little information on the effects of competition in reducing edible yields from carcasses of different size or age, and in different ecological contexts. King (1975), noting most predation and consequently scavenging opportunities to be nocturnal, concludes that diurnally active hominids could not have profitted from scavenging. The conclusion, however, overlooks the possibility of the intensity of competition for kills being variable and at times sufficiently low that nocturnal deaths could persist as scavengeable food sources through the next day. Indeed, aspects of seasonality, habitat type and carcass size that affect carcass persistence are in need of clarification and documentation.

Taphonomists have addressed many of the above anatomically-based deficiencies, but almost exclusively in terms of the bone remains of carcasses whose resource life has long since been exhausted. Most taphonomic research is in these terms physically or mechanistically oriented; carcasses and their component bones are treated as physical objects differentially prone to mechanical forces of dispersal (e.g. Behrensmeyer 1975, A. Hill 1975, 1979, Hill and Behrensmeyer 1985, Voorhies 1969), destruction (e.g. Brain 1967, 1969, 1981, Binford and Bertram 1977), or burial (e.g. Behrensmeyer and Boaz 1980). The mechanistic orientation has provided a good understanding of many of the factors affecting a bone's survival over the largely physical fossilization process.

All bones, however, begin their taphonomic history as structures that bear or contain nutrients actively sought by organisms in need of nourishment. This nutritional or energetic orientation to taphonomy as yet has few subscribers (Binford 1978, Speth 1983). I contend, however, that it is the very process of carcass degradation during the resource life of the carcass that will be most informative about the diet and subsistence of ancient meat-eaters, including the modes by which they acquired meat. In this view, the often dramatic affects of subsequent pre- and post-burial physical taphonomic agents are conditioned largely by the brief and early energetic phases of the taphonomic process. The taphonomic component of a study of scavenging opportunities must therefore be based on observations of fresh carcasses, i.e. recently dead animals, whether whole or partially consumed, that still provide a feeding opportunity to vertebrate carnivores.

Such observations were made during an eleven-month field study in Tanzania's Serengeti National Park and Ngorongoro Crater. Of particular interest was the utilization of fresh carcasses by large carnivores and the resulting scavenging opportunity the carcasses provided. After a description of the field setting and methodology, and the carcass sample observed (Chapter 2), the derivation of two parameters of scavenging opportunities are described (Chapter 3). Chapter 4 contains the main body of field results on the scale and characteristics of modern scavenging opportunity in different ecological and physical contexts. Chapter 5 is an application of these results to the scavenging opportunities potentially encountered by Early Pleistocene hominids of East Africa, and to the archaeological recognition of scavenging.

CHAPTER 2

FIELD STUDIES: THE SETTINGS, METHODS AND SAMPLE

THE SERENGETI AND NGORONGORO ECOSYSTEMS

General Ecology and Settings

The Serengeti National Park and Ngorongoro Crater are tropical savanna-woodland ecosystems located in northern Tanzania (Figure 2.1). Both are protected wildlife refuges administered respectively by the Tanzania National Parks system and the Ngorongoro Conservation Area Authority. The two ecosystems experience a minimum of human activity, restricted mostly to small numbers of tourists, and in the case of Ngorongoro Crater, occasional authorized incursions by Masai pastoralists with their livestock. Both ecosystems support naturally maintained populations of larger mammalian herbivores and carnivores.

The Serengeti: The wildlife and the ecology of the Serengeti have been the subject of intensive study since the late 1950's. Research conducted under the auspices of the Serengeti Wildlife Research Centre (formerly the Serengeti Research Institute) renders the Serengeti one of the world's best known tropical ecosystems. Much of the knowledge acquired is summarized in an excellent volume edited by Sinclair and Norton-Griffiths (1979), to which the reader is referred for a more detailed account of Serengeti ecology than is given below.

The Serengeti National Park has an approximate area of 25000 km^2. It is best known for its vast migratory herds of wildebeest, zebra and Thompson's gazelle, numbering approximately two million individuals. Together with the adjoining Maasi Mara park in Kenya and game reserves, controlled areas and conservation areas that adjoin it in Tanzania, the Serengeti is a self-contained ecosystem with respect to its larger herbivore populations (Sinclair 1979a, Figure 2.1). The migratory herds utilize the woodlands and plains of the Serengeti (Figure 2.2) according to a seasonal pattern of rainfall. As with most of East Africa, the Serengeti experiences a long dry season (June - October), a long rainy season (March - May), and an intermediate period ideally characterized by short rains (November - December) and a short dry period (January - February). In some years, the short rains persist through the short dry season and merge with and intensify into the long rainy season (Sinclair 1979a). The Ngorongoro

Figure 2.1: Northern Tanzania, showing the Serengeti National Park and Ngorongoro Crater (bold lines). The dashed line encompasses the annual range of the Serengeti's migratory herds.

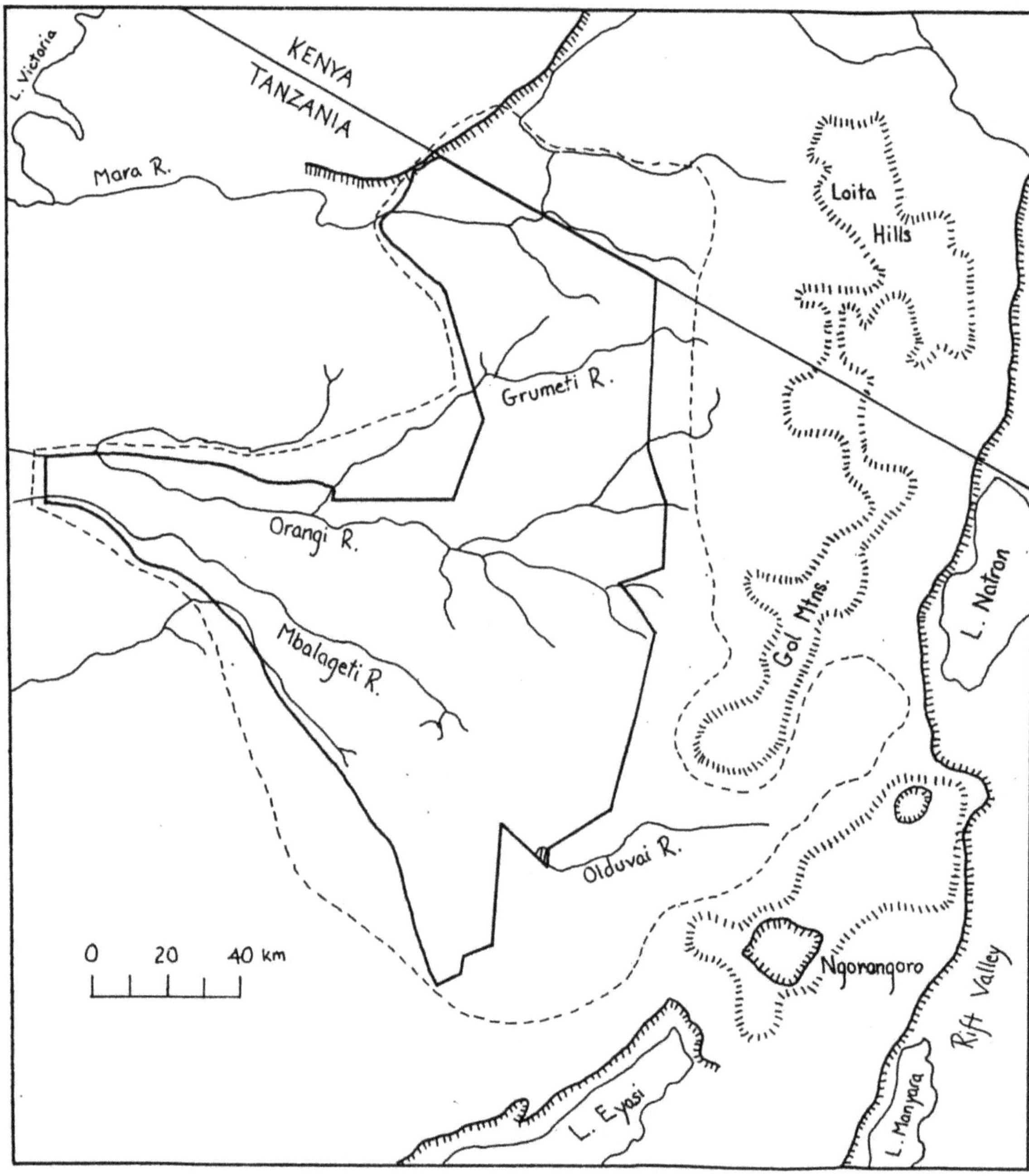

Figure 2.2: The Serengeti National Park, showing main vegetation types and the location of study areas.

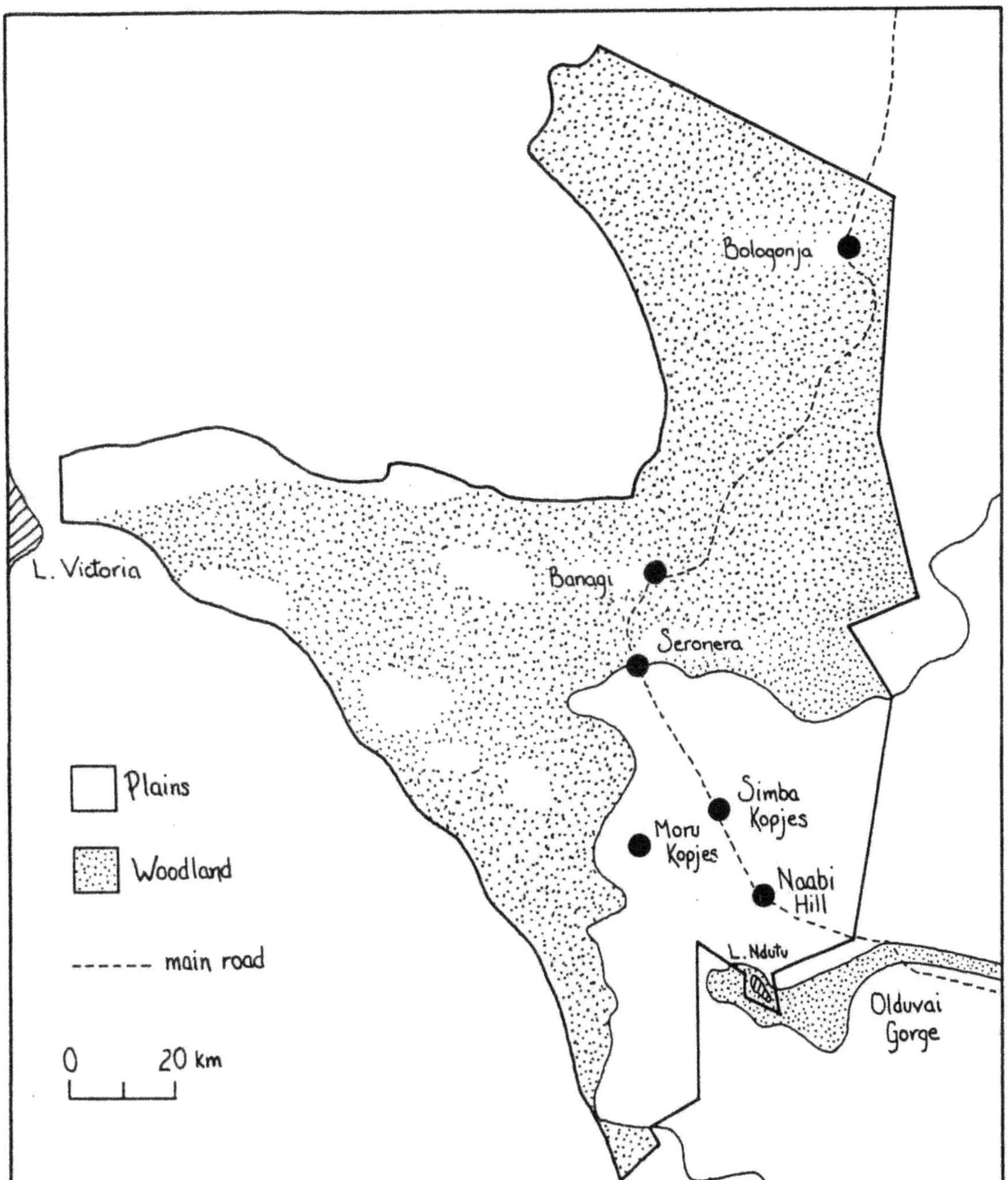

Highlands to the southeast create a rain shadow over the southern plains of the Serengeti, which receives approximately 500mm of rainfall annually. Greater amounts of rain occur as one proceeds north toward Kenya and west toward Lake Victoria out of the rain shadow, reaching a maximum of roughly 1000mm per year (Sinclair 1979a). The average annual rainfall for the Serengeti as a whole is about 800mm (Coe et al 1976).

The migratory herds utilize the Serengeti in a predictable annual pattern (e.g. Pennycuick 1975, Maddock 1979). During the long dry season the herds are dispersed throughout the extensive western and northern woodlands, extending into southern Kenya. With the beginning of the short rains in early November, the herds amass at the southern margin of the woodlands, and, if rain has fallen on the plains, proceed to these prime grazing grounds. The herds remain concentrated on the open plains throughout the short dry and long rainy seasons. With the end of the rains in late May, the herds proceed back to the relatively well-watered woodlands. The seasonal movements of these three migratory ungulate species are a major component of Serengeti ecology (e.g. Bell 1971, McNaughton 1976, 1979, Jarman and Sinclair 1979) due to their great numbers (over 90% of larger herbivores) and biomass (over 60% of larger herbivore biomass) (e.g. Houston 1979).

<u>Ngorongoro Crater</u>: Ngorongoro Crater is a caldera located roughly 40 km southeast of the Serengeti (Figure 2.1). The Crater covers an area of approximately 265 km^2. Its floor lies 1700 meters above sea level, from 400 to 700 meters lower than its surrounding rim (Hay 1976). The Crater floor supports an extensive flat plain surrounding shallow soda Lake Makat, two large swamps, and two small patches of forest (Figure 2.3).

Approximately 900mm of rain falls in the Crater each year (Coe et al 1976) in a seasonal pattern similar to that in the Serengeti. The western portion of the Crater is noticeably drier than the middle and eastern parts, supporting mostly gazelle throughout the year. Dense concentrations of wildebeest and zebra and smaller numbers of buffalo, eland, gazelle, warthog, hartebeest, waterbuck and elephant, to name the most common herbivores, are found mainly in the eastern three-quarters of the Crater. Because of the Crater's relatively high annual rainfall and the availability of permanent fresh water from streams and springs, herbivore numbers fluctuate only to a minor degree on a seasonal basis. During the wet seasons, a very small proportion of Ngorongoro's ungulates (estimated at 9% of all individuals by Frame, 1986) temporarily moves a short distance, less than 10 kilometers, out of the Crater (Elliott and McTaggart-Cowan 1978).

Figure 2.3: Ngorongoro Crater, showing the floor of the caldera. Carcass search was concentrated in the central portion of the Crater, bounded by Lake Makat to the west, Mandusi Swamp and Munge River to the north, and Oljoro Nyuki River and Gorigor Swamp to the east and south.

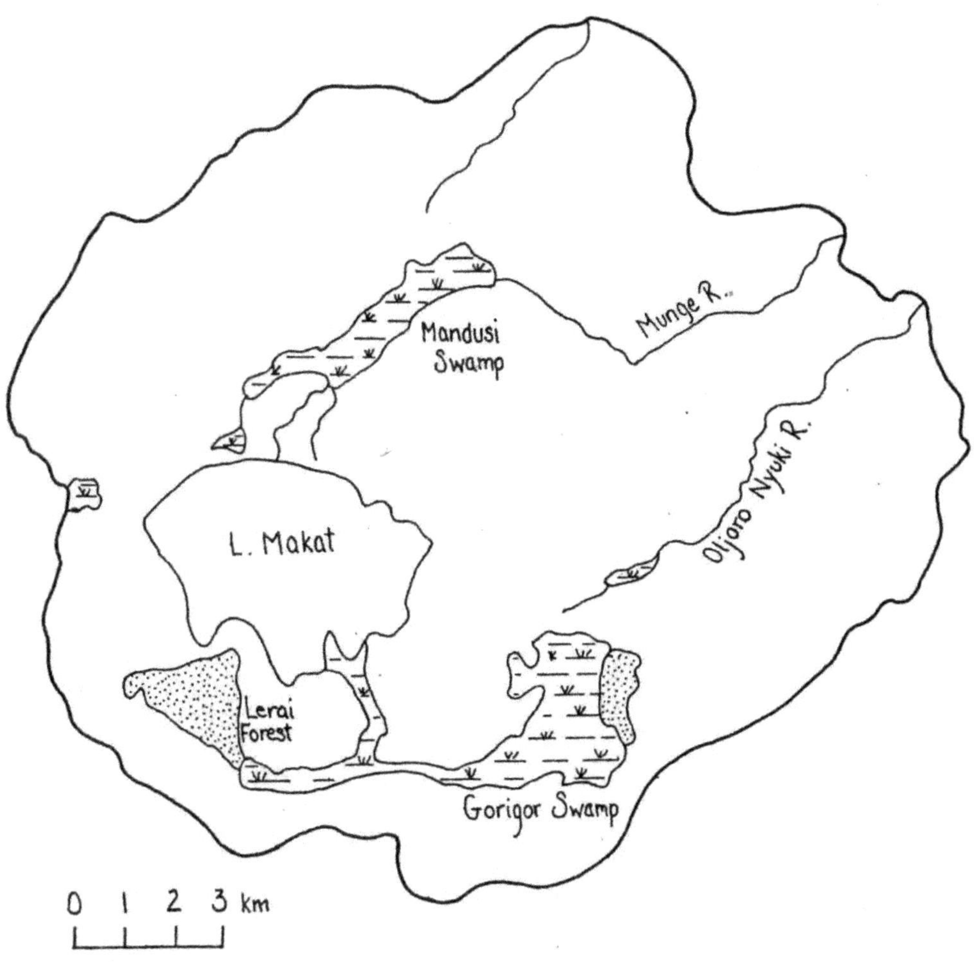

Suitability for the Field Study

Differences in the structure and dynamics of the mammalian herbivore and carnivore guilds between the Serengeti and Ngorongoro and within the Serengeti form the basis for a wide range of scavenging opportunities. This is seen best by comparing mammalian predator to prey ratios, which correspond in a direct manner to the availability of carcasses to scavengers.

Ngorongoro Crater is a relatively stable ecosystem with respect to larger herbivore numbers. As a result, it supports an extremely high density of lions and spotted hyenas (e.g. Elliott and McTaggart-Cowan 1978, Kruuk 1972), among the highest in Africa (C. Packer pers. comm.). Estes and Estes (1979) estimate the ratio of larger mammalian predators (excluding jackals) to ungulate prey to be approximately 1:40. An independent assessment by Frame (1986) agrees well with this figure (1:45 for the dry season, 1:39 for the wet season). A high proportion of herbivore mortality in Ngorongoro results directly from predation, and carcasses can be expected to be subject to complete and rapid consumption by lions and hyenas, both as predators and scavengers. The low availability of carcasses to scavengers in Ngorongoro is attested to by the extremely small number of vultures it supports.

The larger mammalian predator to prey ratio in the Serengeti is up to an order of magnitude lower than that of Ngorongoro. I have calculated the ratio for the Serengeti as a whole to be approximately 1:340 from data presented by Houston (1979) based on census data from the late 60's, and 1:166 from data presented by Frame (1986) based on estimates for 1977. The disparity between these estimates seems to result primarily from a large increase in lion and particularly hyena numbers since the late 1960's, as supported by a May 1977 census on the plains (see Hanby and Bygott 1979 and Sinclair 1979b for explanations of this and other changes in carnivore and herbivore numbers). The relatively small number of carnivores attained by either estimate in comparison to Ngorongoro results from the seasonal migrations of the majority of Serengeti herbivores, and the necessarily fixed annual ranges of its breeding predators (Schaller 1972, Kruuk 1972). Predator populations are therefore largely limited in numbers to those sustainable by the seasonal availability of only small numbers of residential prey (e.g. Sinclair 1979b, Hanby and Bygott 1979). Houston (1979) estimates only 30% of herbivore mortality in the Serengeti to result from predation, ascribing the remainder to natural causes such as disease and starvation. This creates a fundamentally higher availability of carcasses relative to that in Ngorongoro, and corresponds to the large vulture populations existing in the Serengeti (Houston 1972, 1979).

Within the Serengeti itself, the scale of scavenging opportunities can be expected to vary on a seasonal basis dependent on the location of its migratory herds. In areas seasonally occupied by mostly residential herbivores (plains during the long dry season, woodlands during the rainy season) predator to prey ratios will be relatively high and carcass availability to scavengers low. The presence of migratory herds, however, will expectedly increase feeding opportunities for scavengers owing to higher rates of predation and especially more non-predator related mortality. The magnitude of difference can be appreciated through reference to Frame's (1986) data on carnivore and herbivore numbers on the plains during the dry and rainy seasons of 1977. From his data, I calculate the ratio of larger mammalian carnivores (excluding jackals) to ungulate herbivores to be 1:186 during the rainy season, and 1:67 during the dry season. Wet conditions on the plains do result in an estimated quadrupling of larger carnivore numbers, but, due to the more dramatic influx of the major migratory species, a tripling of the predator to prey ratio. Such seasonal changes are not apparent in Ngorongoro, where Frame's (1986) 1977 census data (quoted earlier) indicate seasonal predator to prey ratios to remain steady and perennially high.

The scale of feeding opportunities for scavengers within the Serengeti and compared to Ngorongoro will be discussed fully in Chapter 4. Suffice it to say now that the two ecosystems chosen for the study are ideally suited for revealing a wide range of scavenging opportunities and the characteristics of each.

THE FIELD STUDY

Conditions

The field study was conducted from August 1983 through June 1984. The eleven month period spans the middle of the long dry season to the beginning of the next long dry season. 1983/84 was the second of two relatively dry years in East Africa. The monthly distribution of rainfall in Seronera, on the woodlands-plains ecotone of the Serengeti, is shown for the year encompassing the study period, and is compared in Figure 2.4 to average monthly figures for the preceding ten years. [These data were provided by the Serengeti Wildlife Research Centre, which monitors rain gauges throughout the Serengeti on a monthly basis.] Rainfall amounts for the long dry season (June - October) of 1983 were fairly typical of the preceding ten years, totalling 206.5mm for the five month period in 1983, and an average of 196.8mm during the ten year period. The four month intermediate period between November and February was relatively dry in 1983/84 (246.0mm) compared to the preceding ten years (303.1mm), and the long rains (March -

Figure 2.4: Monthly rainfall (mm) in Seronera (Serengeti Research Institute gauge #35).

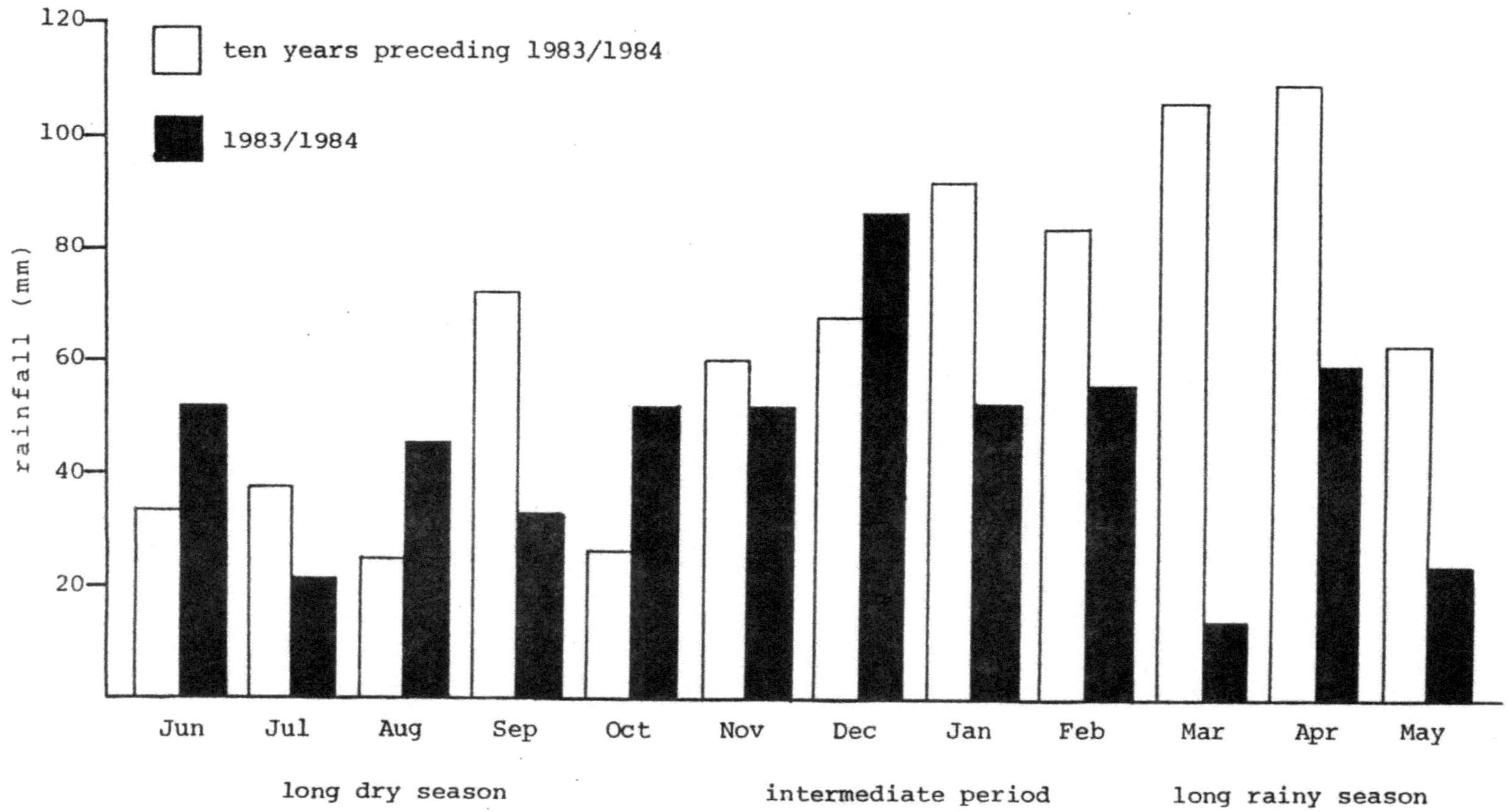

May) all but failed in 1984 (97.6mm compared to the average 278.2mm for the ten year period). Overall, the year encompassing my study experienced over 200mm less rain in Seronera than the average of approximately 780mm for the area. Except for December, the month of my study during which most rain fell in Seronera, rainfall data for 1983/84 show a lack of the pronounced seasonality more typical of Seronera and East Africa in general. The dryness of the study period perhaps resulted in higher than usual herbivore mortality in the Serengeti. This will effect the degree of scavenging opportunity encountered but not the processes and associated characteristics of the opportunity that are of interest. The lack of pronounced seasonality in rainfall, however, denotes that the extent of seasonally based differences in carcass availability was less pronounced than that expected for more typical years. The effect of this will be discussed more fully in Chapter 4.

Study Areas

Characteristics of the study areas, including the intensity by which each was monitored, are given in Table 2.1. The main study area was that around Seronera and my base at the Serengeti Wildlife Research Center. Seronera is located on the woodlands-plains ecotone of the Serengeti (Figure 2.2). The study area includes a diversity of habitats in its roughly 300 km^2 extent, including *Acacia* woodland, plains, and riverine woodland (Figure 2.5). In addition to supporting a diversity of habitat types, Seronera is ideally located with respect to the movements of the migratory herds, a large portion of which pass through the study area on their way to and from the plains. 103 days of carcass searching spanning all eleven months of the study period were spent in Seronera, yielding 150 fresh carcasses, over half of the total carcass sample.

The other major study area within the Serengeti was the Naabi-Ndutu plains (Figure 2.2). Most of this area is a vast, flat open grass plain surrounded by distant hills, but it also includes at its southern end the lacustrine woodlands surrounding perennial Lakes Ndutu and Masek within the western extremity of Olduvai Gorge (Figure 2.6). The Naabi-Ndutu area was sampled systematically for a total of twenty days during the months of January through May when the migratory herds were present in variable and usually very large numbers. Casual searching for carcasses in this area was conducted during periods when only very small numbers of the xerophyllic Grant's and Thompson's gazelle were present. The slight prospects of finding a carcass at these times deemed systematic monitoring unproductive and logistically impractical. 72 carcasses were discovered in the Naabi-Ndutu study area, 32 within or on the margins of the Ndutu woodlands and the remainder on the plains proper.

Table 2.1: Characteristics of study areas in the Serengeti and Ngorongoro.

Study Area	Vegetation	Approx. Area (km^2)	Days of Searching[a]	Months Sampled	Number of Carcasses Discovered
Seronera	woodland, plains, riverine woodland	300	103	Aug-Jun	150[b]
Naabi-Ndutu	plains, lacustrine woodland	400	20	Jan-May	72[c]
Banagi	woodland, riverine woodland	100	7	Aug,Sep Mar-May	5
Simba Kopjes	plains	100	7	Nov,Dec, Feb,Mar, May	14
Incidental (Serengeti)	various	---	3	Jan,Apr, May	5
Ngorongoro	mostly plains	50	25	Nov-Jan, Jun	28

Notes: a) on some days, more than one study area was searched.
b) includes five carnivores.
c) includes one carnivore.

Figure 2.5: The Seronera study area, showing the distribution of the *Acacia* woodland, plains and riverine woodland habitats.

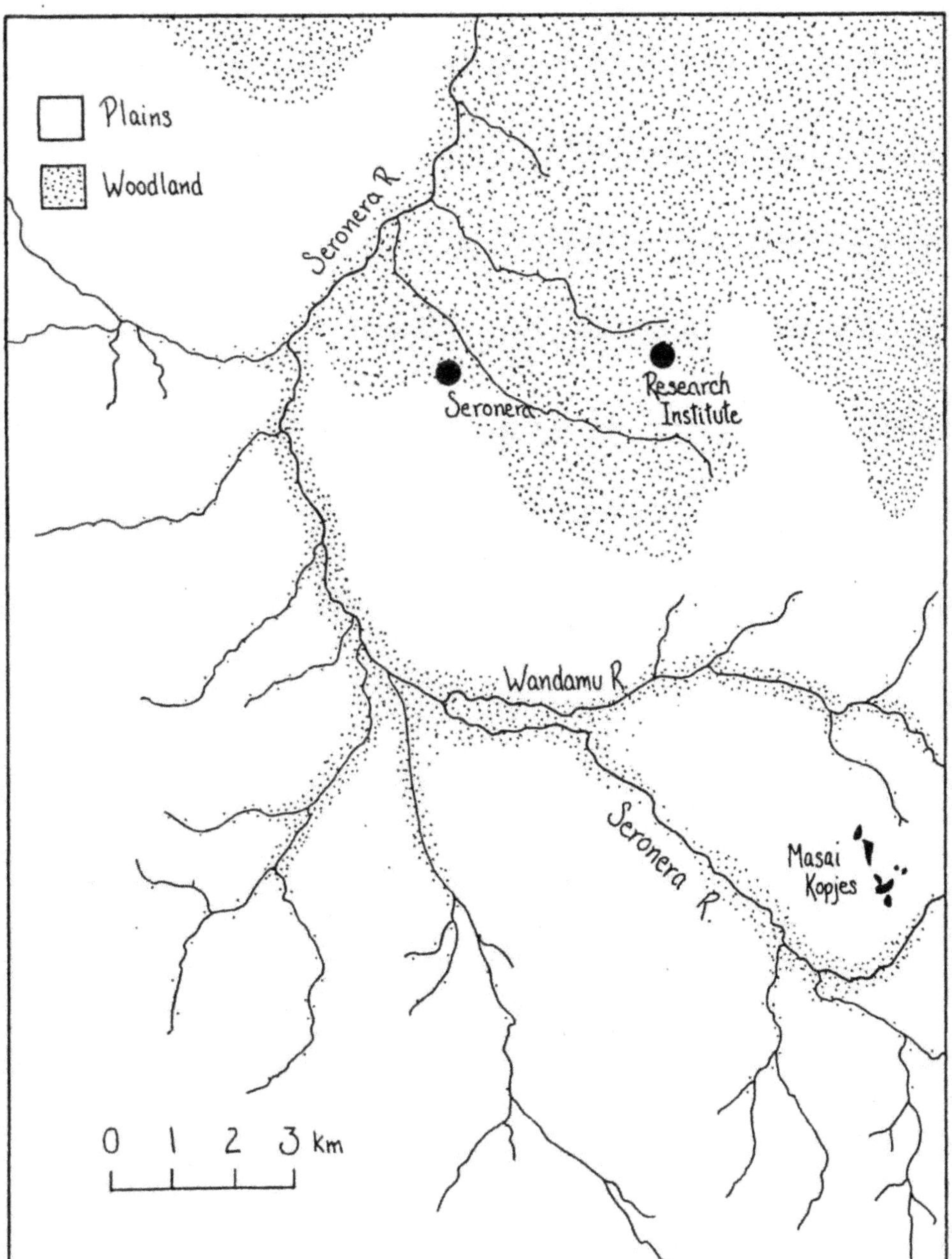

Figure 2.6: The Naabi-Ndutu study area, including the plains to the south of Naabi Hill and the woodlands surrounding Lake Ndutu and Lake Masek.

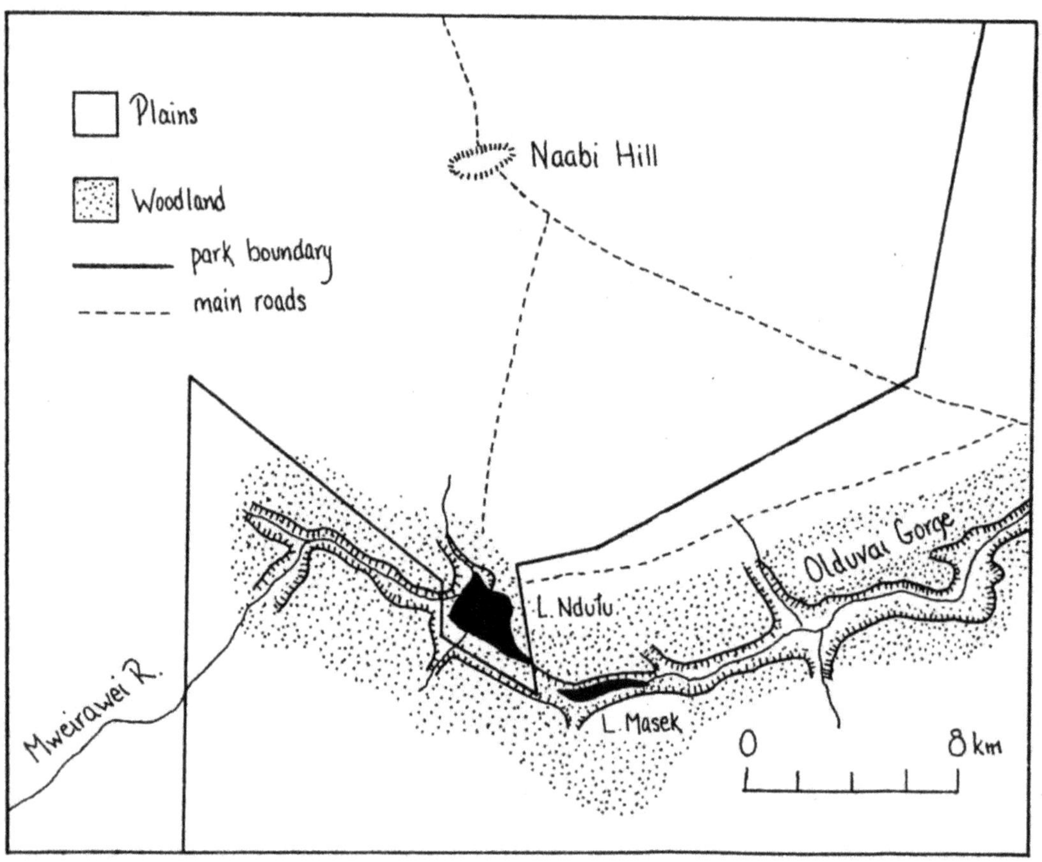

The Banagi and Simba Kopjes study areas (Figure 2.2) of the Serengeti were monitored for only seven days each, mostly during times of migratory prey abundance. Banagi is within the woodlands proper of the Serengeti, characterized by open stands of various *Acacia*, *Commiphora* and other tree species broken occasionally by small patches of lightly treed grassland. Five fresh carcasses were discovered in the area of Banagi. Simba Kopjes is on the Serengeti Plains, punctuated in this area by numerous and substantial kopjes that serve at least seasonally as havens for lions. Fourteen carcasses were observed in this study area.

Five carcasses were discovered incidentally while travelling within the Serengeti. One was found at Bologonja in the northern woodlands, three at Moru Kopjes on the plains, and one on the plains near Olduvai Gorge, on the southeastern park boundary (Figure 2.2).

The study areas in the Serengeti represent a fairly comprehensive sample of Serengeti habitats. However, the western and northern woodlands of the Serengeti were, with the exception of Banagi, not monitored, owing to limited availability of petrol. Therefore, I did not observe the expectedly highest annual mortality that occurs in the northern woodlands at the end of the dry season (Houston 1979). This deficiency, however, is largely compensated for by observations in the Naabi-Ndutu area during late February and March, when unusually dry conditions led to very high levels of natural mortality.

The final study area was in Ngorongoro Crater. 25 days during December, January and June were spent in Ngorongoro, yielding observations on 28 fresh carcasses. Search was concentrated in the wetter and more heavily populated portion of the Crater to the east of Lake Makat, as indicated in Figure 2.3. This is where the majority of the Crater's predation and scavenging occurs.

Field Methods

Carcass search was concentrated during early daylight hours. Search commenced as close to first light as possible and continued throughout the morning or until an area had been satisfactorily covered. Occasional afternoon and evening surveys were made, but with rare exceptions all predator and most other activity had been supressed by heat, and the only carcasses to be found were those already discovered in the morning. Search for and observations at carcasses were not carried out at night, despite the nocturnal peak of predator and scavenger activity. Aside from the logistical difficulties of nighttime observations, my exclusively daytime observations of scavenging opportunity are consistent with those potentially encountered by diurnal prehistoric hominids.

Search was continuous from a slow-moving (25 km/hr.) Land Rover, punctuated by careful, binocular-assisted rooftop scans as frequently as the density of vegetation demanded. The objects of search were concentrations of airborne or grounded vultures, the directed flight of individual vultures, and aggregations of mammalian carnivores. The latter clue to a carcass's presence was the only one available in Ngorongoro.

The abundance of vultures in the Serengeti resulted in their targeting of most carcasses found. By mid-morning (0830-0900) the air is usually sufficiently warm to create thermals on which vultures rise and congregate in large numbers if over a carcass. A large cloud of vultures riding a thermal can be seen from over 10 km away, and was a certain clue to the presence of a carcass below. Vultures provide other less conspicuous, through unambiguous cues to the presence of a carcass. Vultures have difficultly flying without the aid of updrafts. Early morning labored flight of a single vulture was an unmistakable sign of a nearby carcass, and its marked directionality always provided a reliable and easily followed quide to the carcass. An aggregation of as few as five vultures roosting in the lower or middle canopy of a tree was usually a clear sign of a lion kill below. Vultures roosting in small numbers at only the crown of a tree, however, almost always were nesting or occupying a vantage point from which other vultures at or proceeding to a carcass could be located. One quickly learns to recognize these clues for finding carcasses.

Rarely was a carcass found without the clue provided by some carnivore already in attendance. In only a handful of cases was an unattended fresh carcass, whole or partially eaten, discovered. In the case of only two cheetah kills did I follow a predator waiting for a kill to be made. This method, though ensuring the observation of a whole consumption episode, was very inefficient. Indeed, of the over 200 kills eventually discovered, I witnessed the kill on only four occasions. By searching for kills already made, I maximized the coverage of an area for all carcasses present or potentially so.

Although originally proposed, search was not conducted along systematic transects (e.g. Anderson and Pospahala 1970, Burnham and Anderson 1976, Eberhardt 1978, Robinette et al 1954, 1974). Several early attempts with this method proved to be unproductive and an extravagant use of petrol. Carcasses in tropical ecosystems have a very short persistence compared to those in temperate latitudes, where, for example, a winter's worth of deer mortality can be estimated in one series of springtime transects (e.g. Robinette et al 1954). Rather, search was much more selective, being concentrated when possible in areas where I knew predators to be present. This method was particularly useful in locating lion kills in dense vegetation or carcasses during very early daylight hours. Alternatively, areas selected for search were those in the vicinity of the larger herds of ungulates which were typically patchily distributed in a study area. The method adopted was appropriate, since I was more interested in the variable feeding potential provided by individual carcasses than absolute, instantaneous carcass densities.

Most carcasses being consumed could be observed within ten meters if approached slowly and indirectly. This distance was adequate for making detailed, binocular-assisted observations of the carcass parts being consumed. Carnivores were extremely tolerant of my presence and ginger maneuvering for a better view, especially in the study areas most intensively monitored. In only a handful of cases did my presence noticeably disturb a consumer, and in only one case of a baboon feeding on a Thompson's gazelle fawn did I interfere with the normal pattern of carcass consumption.

A number of records were made consistently upon discovery of a carcass. These include the habitat, approximate location, the types and estimated number of herbivores in the area, the number and types of actual consumers and potential scavengers present, and the distribution and amount of edible tissues on the carcass. All observations were made from within the vehicle if consumers were present, but if not, a much more detailed, out-of-vehicle inspection of the carcass could be made. Whether conducted on discovery or after abandonment by consumers present on discovery, these inspections additionally included a determination of the sex, dental age and, when possible, the quality of the carcass's marrow. Marrow quality was determined using visual criteria (e.g. Brooks et al 1977, Franzman and Arneson 1976, Harris 1945, Ransom 1965, Sinclair and Duncan 1972).

Observations on carcass consumption included a minute-by-minute account of the parts being consumed or removed from the carcass, and the identity and number of consumers. Unless immediately scavenged by hyenas, a detailed inspection of the carcass was made upon abandonment by successive consumers. I attempted whenever possible to make continuous feeding observations until all edible tissues had been consumed. For logistical reasons such as extremely prolonged persistence of a carcass, this goal could not always be achieved. In many cases, as well, slow and minimal feeding and the presence of another carcass in the area inhibited continuous observations. Efforts were always made to revisit carcasses and periodically monitor their resource status until they no longer represented a source of scavengeable food. I was much more successful at this in the intensively monitored Seronera and Ngorongoro study areas, than in other study areas which were usually visited for only one or several days at a time.

THE CARCASS SAMPLE

Observations were made on the consumption and scavenging potential of 274 fresh carcasses. Again, a fresh carcass is defined as any recently dead larger mammal still providing nourishment to vertebrate carnivores. Fresh carcasses are in this way clearly distinguished from weathered bones of carcasses whose resource life has been exhausted and which occur in far greater numbers on the landscape. From hereon, the term carcass will be used to mean exclusively a fresh carcass. Six mammalian carnivores are included in the total carcass sample, including one spotted hyena, two cheetah, two jackals, and one bat-eared fox. The carnivores are excluded from the following analyses, so that the total sample of larger mammalian herbivore carcasses numbers 268. This large sample of fresh carcasses permits for the first time a detailed assessment of the many factors affecting scavenging opportunities.

The taxonomic composition of the carcass samples from the Serengeti and Ngorongoro are shown in Tables 2.2 and 2.3, respectively. Also presented in each table are the late 1960's and late 1970's sets of comprehensive population size estimates of the more common larger mammalian herbivores provided by Houston (1979), Elliott and McTaggart-Cowan (1978), and Frame (1986). The data presented for Ngorongoro indicate a marked degree of stability in herbivore numbers, at least during the penultimate decade. The live population size estimates for the Serengeti, on the other hand, indicate some dramatic changes for certain species during the 70's. For example, there had been a doubling of the number of Serengeti wildebeest by the late 70's (see also Sinclair 1979b), and unpublished data collected by Marcus Borner indicate the Thompson's gazelle population to have continued its decline of the 70's to a current estimated number of only 250,000. With these temporal differences in mind, the available live population size estimates are still useful for making broad species-specific comparisons with carcass population sizes.

Wildebeest are by far the dominant species in the carcass sample. Wildebeest represent 60% and 82% of all carcasses from the Serengeti and Ngorongoro, respectively. Their dominance results in part from their great live abundance in both ecosystems, and additionally from their status as the preferred, easily caught prey of lions and hyenas (Elliott et al 1977, Kruuk 1972, Schaller 1972). Zebra are the next best represented species in the Serengeti carcass sample, occurring, like wildebeest, more frequently than expected from their relative live abundance. Thompson's gazelle, a small antelope, are, conversely, uncommon in the Serengeti carcass sample relative to their great live abundance (although here, a current live population size of 250,000 would lower their relative live abundance to a level approximating their

Table 2.2: Species composition and relative abundance of the Serengeti carcass sample and live herbivore community. Only more common live herbivores are included, which account for more than 99.5% of the Serengeti's total larger herbivores (Houston 1979). Taxa are ordered according to increasing weight of adults.

	Carcass Sample		Estimated Live Abundance: Houston (1979), for late 1960's		Estimated Live Abundance: Frame (1986), for 1977	
Species	n	%	n	%	n	%
Thompson's gazelle	24	10.0	981000	42.0	650000	22.1
Grant's gazelle	6	2.5	6000	0.3	52000	1.8
Impala	3	1.3	119100	5.1	221920	7.6
Warthog	1	0.4	34200	1.5	33330	1.1
Topi	2	0.8	55500	2.4	172160	5.9
Wildebeest	144	60.0	720000	30.8	1440000	49.1
Hartebeest	0	0.0	20700	0.9	26445	0.9
Waterbuck	1	0.4	3000	0.1	3540	0.1
Zebra	49	20.4	240000	10.3	200000	6.8
Eland	0	0.0	24000	1.0	28180	1.0
Buffalo	6	2.5	108000	4.6	75000	2.6
Giraffe	2	0.8	17400	0.7	26270	0.9
Rhinoceros	1	0.4	900	0.04	1500	0.05
Hippopotamus	0	0.0	2400	0.1	500	0.02
Elephant	1	0.4	4500	0.2	3850	0.1
TOTAL	240	99.9	2336700	100.04	2934695	100.1

Table 2.3: Species composition of the Ngorongoro carcass sample and live herbivore community. Only the more common larger herbivores are included in the table. Taxa are ordered according to increasing live adult weight.

Species	Carcass Sample		Estimated Live Abundance: Elliott + McT.-Cowan (1978), for 1969-71		Estimated Live Abundance: Frame (1986), for 1977	
	n	%	n	%	n	%
Thompson's gazelle	0	0.0	3750	13.0	3657	13.6
Grant's gazelle	0	0.0	1700	5.9	1578	5.9
Wildebeest	23	82.1	17000	58.8	16535	61.5
Hartebeest	0	0.0	150	0.5	145	0.5
Waterbuck	0	0.0	125	0.4	31	0.1
Zebra	2	7.1	5500	19.0	4499	16.7
Eland	0	0.0	475	1.6	214	0.8
Buffalo	3	10.7	200*	0.7	228	0.8
TOTAL	28	99.9	28900	99.9	26887	100.1

* my estimate of those present in June 1984.

relative abundance as carcasses). Smaller taxa in general are underrepresented as carcasses in both ecosystems, and especially in Ngorongoro. This is clearly not due to lower mortality rates, but rather to the rapidity by which they are consumed relative to larger animals (see Chapter 4). The relative abundance of species in the carcass sample is therefore not representative of total mortality, but is rather a skewed profile corresponding more closely to the likelihood of a scavenger encountering any one species as a carcass. This point will be returned to at the end of Chapter 4.

A number of taxa occurring in the two ecosystems are not represented in the carcass sample. These species in the Serengeti (hartebeest, eland, hippopotamus) are relatively uncommon, though regularly seen in their preferred habitats. More common taxa poorly represented in the carcass sample are all smaller species under fifty kilograms live adult weight.

The size and age composition of the total larger herbivore carcass sample is presented in Table 2.4. The size class designation for each taxon corresponds to that commonly used by archaeological faunal analysts (e.g. Brain 1981, Bunn 1982), with the important exception that young individuals of a particular taxon are placed in a smaller size category than adults. This is necessary because large developmental size changes are expected to result in different treatment by consumers of young and adult carcasses of the same species. The age classes used correspond to the dental age assignments used by archaeological faunal analysts (e.g. Bunn 1982): neonates and all pre-juveniles (Bunn's Age 1) display at most slightly worn deciduous premolars, with at most the first of the molars erupting. Juveniles (Bunn's Age 2) have permanent premolars, but only slightly worn permanent molars and incompletely erupted third molars. Adults (Bunn's Age 3 and Age 4) display a full complement of permanent dentition, all of which is in wear.

The sample has been stratified into six size/age categories. The carcasses comprising each category are shown in Table 2.5. The number of categories created reflects an attempt to maximize the sample size while minimizing the size and age range of each. The categories comprise the basic carcass analytical units to be used throughout the presentation of results in the following chapters.

Two subadult categories have been created. The "small subadult" category is dominated by Thompson's gazelle fawns and juveniles and neonate wildebeest; it contains 34 carcasses, or 12.7% of the total carcass sample. The "medium-sized juvenile" category is comprised exclusively of juvenile wildebeest and zebra, with the exception of one buffalo calf. 28 carcasses, or 10.5% of the total sample,

Table 2.4: Size and age composition of the total larger mammalian herbivore carcass sample. For the age column, N = neonate or pre-juvenile, J = juveniles, SA = all subadults, A = adults, T = total. Bunn's (1982) size groups are those commonly used by archaeological faunal analysts. The size designations used here take account of large developmental size changes in any one species.

Species	Age	Adult Weight (kg)	Bunn's Size Group	My Size Group	Carcass Sample n	%
Thompson's gazelle	SA		1	1	11	4.1
	A	15	1	1	13	4.9
	T				24	9.0
Grant's gazelle	N		2	1	3	1.1
	A	40	2	2	3	1.1
	T				6	2.2
Impala	J		2	1	1	0.4
	A	40	2	2	2	0.7
	T				3	1.1
Warthog	A/T	45	2	2	1	0.4
Topi	N		3	1	1	0.4
	A	100	3	3	1	0.4
	T				2	0.7
Wildebeest	N		3	1	15	5.6
	J		3	2	14	5.2
	A	123	3	3	137	51.3
	T				166*	62.2
Waterbuck	J/T	160	3	2	1	0.4
Zebra	N		3	1	4	1.5
	J		3	2	11	4.1
	A	200	3	3	36	13.5
	T				51	19.1
Buffalo	N		4	2	1	0.4
	A	450	4	4	8	3.0
	T				9	3.4
Giraffe	A/T	750	5	5	2	0.7
Rhinoceros	A/T	816	5	5	1	0.4
Elephant	A/T	1725	6	6	1	0.4
TOTAL	SA				62	23.2
	A	---	--	--	205	76.8
	T				267	100.0

* excludes 1 wildebeest carcass for which age could not be determined.

Table 2.5: Size and age stratification of the total carcass sample. The size/age categories, based on data in Table 2.4, are the basic carcass analytical units to be used throughout.

Carcass Size/Age Groups	Species and Age Composition (n)	Subtotal n/(%)
Small subadults	Thompson's gazelle subadults (11) Grant's gazelle neonate (3) Topi neonate (1) Wildebeest neonates (15) Zebra neonates (4)	34 (12.7)
Small adults	Thompson's gazelle (13) Grant's gazelle (3) Impala (2*) Warthog (1)	19 (7.1)
Medium-sized juveniles	Impala (1*) Wildebeest (4) Waterbuck (1) Zebra (11) Buffalo neonate (1)	28 (10.5)
Medium-sized adults	Topi (1) Wildebeest (137) Zebra (36)	174 (65.2)
Large adults	Buffalo (8)	8 (3.0)
Very large adults	Giraffe (2) Rhinoceros (1*) Elephant (1)	4 (1.5)
TOTAL		267

* indicates poached individuals. The two impala were snared and discovered before poachers, while poachers had severely interfered with the rhino before my discovery of it; the rhino is excluded from the following analyses.

are classified as medium-sized juveniles.

Four adult size categories have been created. Size 1 and Size 2 adults are lumped into a "small adult" category to create a sufficiently large sample of 19 carcasses, or 7.1% of the total. The small adult category is dominated by Thompson's gazelle, and also includes adult Grant's gazelle and impala. "Medium-sized adults" are by far the largest category, including 174 carcasses or 65% of the total sample. Adult wildebeest dominate this category, while the remainder of the carcasses are all zebra adults with the exception of one adult topi. "Large adults" are represented exclusively by eight buffalo, comprising 3% of the total sample. "Very large adults" include only four carcasses, one of which (the rhinoceros) was poached and too severely interfered with by poachers to contribute to the following analyses. The very large adult category is therefore represented by two giraffe and one elephant, less than 2% of the total sample.

Most of the six size/age categories are sufficiently large for evaluating general size- and age-based differences in carcass utilization. Only the medium-sized adult category, however, contains enough individuals to permit many of the detailed, multiple variable analyses presented in Chapter 4.

CONSUMERS OBSERVED

Lions, spotted hyenas and vultures were the most frequently observed carcass consumers. Other consumers observed were cheetah, jackals (three species), bat-eared foxes, baboons, wild dogs, maribou storks, and tawny eagles. These other consumers were either infrequently observed at carcasses, or, as with jackals and maribou storks, were commonly present but made immeasurable contributions to carcass consumption.

Leopards and striped hyenas occur in the Serengeti in very small numbers. These species were seen on only two occasions each and never at a carcass. The term hyena therefore will be used throughout to refer exclusively to the spotted hyena. A diversity of smaller mammalian carnivores occurs in the Serengeti and Ngorongoro, but these carnivores are mainly nocturnal species never observed at carcasses.

The term vulture is used in a generic sense throughout. The vulture taxa most frequently present at carcasses and in the largest numbers are the griffon vultures, including the white-backed vulture and Ruppell's griffon vulture. Other vulture species frequently present at carcasses but always in small numbers are the lappet-faced vulture and the hooded vulture. The Egyptian vulture was rarely observed at carcasses in the Serengeti, but was present on a number of occasions singly or in pairs in Ngorongoro.

CHAPTER 3

PARAMETERS OF SCAVENGING OPPORTUNITY

I am concerned here with the development of two parameters of scavenging opportunity. These are the completeness of carcasses available to scavengers, and the duration over which carcasses persist as scavengeable food sources. The bulk of discussion concerns an analysis of the sequences by which carcass parts are consumed, the results of which are critical for establishing a meaningful measure of carcass completeness. The two parameters are used in the following chapter in an analysis of variations in the scale of scavenging opportunity among particular physical and ecological contexts.

CARCASS COMPLETENESS

The completeness of a carcass determines the types and amount of edible tissues available to scavengers. Scavengers typically encounter incomplete carcasses; upon discovery, 258 of 268 carcasses in my sample had been or were being consumed. The less complete this initial consumption, the greater the diversity of parts available to a scavenger, the greater its total nutritional reward, and, in general, the greater the scavenging opportunity.

For my purposes, a measure of carcass completeness must be informative on two fronts. First, the measure must reflect the fraction of a carcass's resource potential remaining for scavengers. Second, the measure must be sensitive to the specific carcass parts comprising the feeding opportunity, and in particular the series of skeletal elements bearing the remaining edible tissues. The latter requirement is essential to the ultimate application of observations on modern carcass use and availability to those feeding opportunities a hominid scavenger might have encountered, and to the development of criteria for the archaeological recognition of scavenging.

The ability of a measure of completeness to characterize accurately both the amount and distribution of edible carcass parts depends on the degree of uniformity in carcass consumption patterns. An invariant sequence of carcass part consumption by all carnivore taxa on all carcass types would render as highly specific the yield and series of carcass parts represented by a single completeness value. A random or inconsistent consumption pattern, on the other hand, would render the value insensitive to the parts comprising it, and consequently of little use for archaeological predictions. To evaluate the extent to which one can accurately speak of a general consumption pattern, data are presented below on the sequence by which various carnivores have been observed to consume the parts of variously sized and aged carcasses. Although presented here for methodological purposes, it will be seen in Chapter 5 that the consumption sequence itself provides a basis for criteria for distinguishing archaeological assemblages accumulated through scavenging as opposed to hunting.

Carcass Consumption Sequences

Constructing A Sequence: Carcass consumption sequences are constructed through a cross-sectional analysis of the part by part completeness of all carcasses. Per-part completeness is that encountered upon my discovery of a carcass. These data were collected whether or not the carcass was being consumed and regardless of the stage consumption had reached in a way described fully in Chapter 2. On-discovery data are used in order to ensure the best possible discrimination among those parts typically consumed early in the sequence, and for which feeding observations are the least frequent. In a number of cases I was unable to appraise accurately the completeness of all parts on discovery, so that data from the earliest systematic inspection conducted between feeding bouts of successive consumers are used. For only 8 of 268 carcasses discovered did circumstances prevent an accurate and simultaneous assessment of completeness at some point before the carcass's resource life had been exhausted. Most uncertainties were restricted to the completeness of only one or two parts, and these were usually unambiguously cleared up by consulting records on subsequent feeding observations or systematic inspections.

Field records on carcass completeness on discovery were encoded for computer analysis by assigning a whole number completeness value to each of a series of 26 carcass units. The 26 parts constitute the whole of a carcass minus the tail and body skin (see, for example Table 3.1). With the exception of the lower viscera (digestive organs) and upper viscera (heart, lungs, diaphram, esophagus), each unit corresponds to the bone or bone set that bears or contains the edible tissue.

The carcass parts used were usually easily distinguished during feeding observations. This is certainly true for tissues contained by bone (limb bone marrow, brain and head pulps), and for most flesh parts as well. Slight ambiguity arose only in distinguishing pelvic girdle flesh from both proximal femur flesh and posterior lumbar flesh.

The specificity of the parts used corresponds to the manner by which carcasses are consumed by carnivores. Rib flesh and thoracic vertebrae flesh are considered as one unit, the ribcage (unlike Binford, e.g. 1978). This is because carnivores do not, and I was unable to, distinguish between proximal ribs and thoracic vertebrae during the consumption of dorsal ribcage flesh. For similar reasons (and again unlike Binford), sacral flesh is not distinguished from pelvic flesh (being labeled pelvic flesh), nor are the tarsal and carpal bones distinguished from the metapodials (the unit being referred to as either metatarsals or metacarpals).

The average rank in the consumption sequence of paired, appendicular parts is used. Upper and middle limb bones (humerus, femur, radius-ulna, tibia), however, are each considered twice to account for and distinguish between flesh borne by the parts and the marrow contained by them. The flesh of a given limb bone is typically eaten much earlier than its marrow. Foot bones (metapodials and phalanges), however, bear negligible to no flesh an are considered only once, for their marrow. The head, like upper limb bones, bears flesh and contains edible tissues, so that a distinction is made between head flesh parts (tongue, mandible, maxilla) and head contents (brain, and nasal, mandibular and frontal pulp). Postcranial axial units, on the other hand, are considered only for their flesh, as the quantity of grease in these bones is small and could not have been extracted by hominids before the late advent of fire for cooking and containers in which grease could be rendered.

A whole number value of 0 to 4, ranging from least to most complete, was assigned to each unit of all carcasses, as follows:

- 0 = the edible tissue is finished, and the associated bony part is either present or absent.
- 1 = the edible tissue is almost finished, with less than approximately 25% remaining.
- 2 = the edible tissue is approximately half eaten.
- 3 = the edible tissue has been started, but is greater than approximately 75% complete.
- 4 = the edible tissue is complete.

Additionally, if the completeness of a carcass part was never accurately assessed, the part was assigned a value of 9 and excluded from subsequent analysis.

Five grades of per-part completeness were used in recognition of repeated observations that many parts are, to varying degrees, incompletely consumed before feeding commences or even is finished on other parts. The use of a simpler presence/absence or three tier coding system would obscure this important tendency of scavengers to encounter partially consumed parts.

The completeness values for like parts among a sample of carcasses were analyzed in a cross-sectional manner. For each part a mean completeness value and a frequency distribution of values were obtained as alternative bases for ranking parts into a consumption sequence. The cross-sectional method assumes the state on discovery of each carcass to be representative of a discreet stage in the consumption sequence. The method has the advantage over the use of longitudinal feeding observations at a series of individual carcasses of enabling the use of all carcasses rather than a much smaller sample for which the carcasses' whole or nearly whole resource life was observed. The major disadvantage of the cross-sectional method is its requirement of relatively large sample sizes to obtain an adequate level of discrimination among parts of similar rank in the consumption sequence.

Mean completeness values are used to rank carcass parts into a consumption sequence. A part with a mean completeness value of, for example, 1.6 is eaten on average earlier than a part with a mean completeness value of 1.9. Parts with equal mean completeness values are eaten on average simultaneously and given equal rank values. The meaning of these values, however, is ambiguous. The ranking neither refers to the sequence by which parts are started (as would be the case if the ranking was based on the frequency by which parts were encountered complete), nor to the sequence by which parts are finished (as would result from the use of frequency values of parts encountered finished). Rather, the ranking based on mean values indicates a sequence by which parts are partially eaten, which has the advantage over other methods of being sensitive to the common partial use of parts that are either abandoned as such, or not completely consumed until later in the sequence. This is particularly the case with nasal pulp, the anterior portion of which, owing to the softness of the nose, is typically started very early in the sequence, but not finished until more attractive within-bone tissues are consumed (Table 3.1). It is also true for the tongue, which is typically started after mandible flesh, but rapidly finished before remaining head flesh is completely consumed (Table 3.1). Minor cases occur with humerus flesh, metatarsal marrow and the brain (Table 3.1). With these few and usually minor exceptions, the sequences based on mean, frequency-complete and frequency-finished values are very similar, with the mean value ranking being intermediate in cases where they differ (Table 3.1).

It should be noted that the ordering of carcass parts in Table 3.1 and similar tables is based on their anatomical proximity to one another, rather than strictly by their actual rank in the consumption sequence. This arrangement was deemed easier to interpret than if parts were ordered irrespective of anatomy to obtain a perfectly correlated relationship with rank.

The General Carcass Consumption Sequence: The consumption sequence for all 260 herbivore carcasses for which data were available is shown in Table 3.1. The sequence is a general consumption sequence (G.C.S.) because it disregards differences in carcass and consumer type.

Edible tissues occurring outside of bones (viscera and flesh) are consumed before tissues within bones (marrow and head contents). Within these two tissue categories, carcasses are consumed from the back to the front among three distinct body regions. Six stages of carcass consumption, listed from the earliest to last eaten, are recognized:

1) hindquarter flesh, including viscera and flesh posterior to the ribcage, with the exception of tibia flesh, which is consumed in the next stage.
2) forequarter flesh, i.e. post-cranial flesh anterior to and including ribcage flesh.
3) head flesh, including the tongue.
4) hindlimb marrow.
5) forelimb marrow.
6) head contents.

With the exception of tibia flesh, all parts within any one of the six body regions are completely consumed before parts of the region begun next in the G.C.S. The exception provided by tibia flesh results from the late consumption of the distal half of the unit, which is tightly encased in skin and eaten only after more easily extracted flesh parts.

Expected Deviations from the G.C.S.: The possibility of consumer- and carcass-specific deviations from the G.C.S. has important archaeological consequences, as mentioned earlier. Field observations and theoretical considerations indicate that consumption patterns should vary according to two major factors: 1) different mechanical constraints imposed on the consumption of carcasses of different size, and the differential ability of consumer taxa to overcome these constraints, and 2) variations in the relative nutritional attractiveness of parts related to carcass age differences. Dividing the total sample into individual carcass and consumer categories reduces sample sizes to the extent that it is not always possible to obtain sufficient discrimination among parts in all potentially relevant consumer and carcass groupings. Nonetheless, consumption sequences for a number of such groupings could be established, as shown in Tables 3.2 - 3.4.

Table 3.2 shows the effect of carcass size on the consumption sequence. Carcass size is an expected basis for

Table 3.1: The consumption sequence for all carcasses (n=260) based on a ranking of 1) mean values of per part completeness on discovery of the carcasses, 2) the frequency by which parts were discovered complete, and 3) the frequency by which parts were discovered finished. Note the general similarity among the three sequences, with major exceptions being restricted to the rapidity by which the tongue is finished (a), and the early initiation of feeding on nasal pulp (b). The rank based on mean values is used for the general consumption sequence (G.C.S.).

Carcass Unit	RANK IN CONSUMPTION SEQUENCE mean values	frequency complete	frequency finished
lower viscera	1	1	1
upper viscera	4	5	4
HINDQUARTER FLESH			
pelvis	2	2	3
lumbar	5	4	5
femur	3	3	2
tibia	9	9	9
FOREQUARTER FLESH			
ribcage	6	6	8
scapula	8	7	7
humerus	7	8	6
radius-ulna	11	11	12
cervical	10	10	11
HEAD FLESH			
tongue	12	13	10[a]
mandible	13	12	13
maxilla	14	14	14
HINDLIMB MARROW			
femur	15	16	15
tibia	16	17	16
metatarsal	18	20	18
phalanges	17	19	17
FORELIMB MARROW			
humerus	19	18	19
radius	21	21	21
metacarpal	22	23	22
phalanges	20	22	20
HEAD CONTENTS			
nasal pulp	23	15[b]	24
mandible pulp	25	24	25
brain	24	25	23
frontal pulp	26	26	26

deviations from the G.C.S. because large carcasses potentially contain parts that pose physical constraints to consumption not encountered on smaller carcasses. Only adult carcasses are considered in order to control possible age-related variations. Consumer type, however, could not be controlled without reducing the sample sizes of the small adult and large adult classes to inadequately small numbers. Medium-sized adults (wildebeest and zebra) are by far the largest sample (n = 168 carcasses) and as a result show the greatest similarity to the G.C.S. Large adults (buffalo) are represented by only eight individuals, resulting in poor rank discrimination of parts eaten late in the sequence (marrow and head flesh). This problem is not seen with the small adult sample (mostly Thompson's and Grant's gazelle), which numbers 24 carcasses. Very large adult carcasses (giraffe and elephant) are excluded from the table due to small sample size (n = 3).

The effect of carcass age on the consumption sequence is shown in Table 3.3. Carcass age is an expected basis for deviations from the G.C.S. primarily because of age-related differences in the nutritional content of marrow bones relative to other parts. Marrow cavities of neonate and young juvenile ungulates function exclusively as sites for the production of red blood cells and therefore contain only hematopoietic fluid (e.g. deCalesta et al. 1977, Hanks et al. 1976, personal observations). Marrow cavities of adults, however, function primarily as fat reservoirs for use by the animal during nutritionally stressful periods (Brooks 1978, Brooks et al. 1977, Dunham and Murray 1982). The transition in marrow cavity function occurs during the juvenile period (personal observation), such that marrow bones of carcasses designated juvenile contain variable proportions of hematopoietic fluid and fat. Because fat has a much higher nutritional yield, one might expect marrow bones of young animals to be ignored until consumption is completed on other parts ranked later in the adult-dominated G.C.S. The consumption sequences for carcasses representing neonates, juveniles and adults are shown in Table 3.3. Carcass size within any one age class is controlled by considering the three developmental stages of wildebeest and zebra only. Consumer type could not be held constant without severely reducing sample sizes.

Consumer type is the final major factor expected to effect variations from the G.C.S. The greatest difference is expected in the differential ability of carnivores to gain access to tissues within bones. Hyenas are noted for their ability to expose nutrient-rich bone cavities in even the largest animals, and are the primary consumers of marrow and head contents in Africa. The less powerful jaws of felids permit only limited access to these tissues, and all vulture species (with the exception of the lammergeier, a rare species which drops bones from heights onto rocks) are completely unable to do so. Carnivores able easily to extract within-bone nutrients might be expected to consume these parts at an earlier stage than predicted by the G.C.S., which is composed of carcasses eaten by all carnivore taxa observed.

The consumption sequences for lions, hyenas and vultures on medium-sized adult carcasses is shown in Table 3.4. Only those carcasses are included whose context and condition (e.g. degree of bone modification and disarticulation) permitted a high level of certainty that only one identifiable consumer taxon had fed. The consumer specific analysis is necessarily limited to the large sample of medium-sized adult carcasses. Sample sizes for other age and size classes were too small to yield useful results. Incidental comments, however, will be made on these other carcass types, and on the cheetah, a not rarely observed consumer restricted to small prey.

Deviations from the G.C.S.: The specific consumption sequences shown in Tables 3.2 - 3.4 are all broadly similar to the G.C.S. in Table 3.1. All flesh parts, with the occasional exception of flesh on the anterior portion of the body, are consumed before marrow and head contents. Numerous deviations occur among individual carcass parts within any of the six body regions outlined for the G.C.S. These are usually slight and are probably due largely to sampling and statistical limitations. The notable deviations from the G.C.S. are those involving a partial or whole sequential switch between adjacent members of the six body regions, as discussed below.

The three flesh bearing regions are eaten among all consumer and carcass categories in a pattern very consistent with the G.C.S. The major deviations are the consumption of head flesh before some forequarter flesh parts. Vultures are late consumers of most forequarter flesh relative to head flesh on medium-sized adults (Table 3.4), and, it was observed, on all carcasses in general. This results from the limited ability of vultures to gain access to flesh beneath skin except from natural orifices, including the anus, genitals, ears, eyes, and mouth. Parts close to orifices (head flesh, pelvic, lumbar and femur flesh) are therefore consumed by vultures before hidden tissues (upper viscera, forelimb, tibia and cervical flesh). Head flesh is also consumed before cervical flesh, and to a lesser extent radius-ulna flesh, in a number of contexts other than vultures feeding alone. Cervical flesh on large adult (buffalo) carcasses is covered with particularly thick skin which inhibits access to the anterior portion of the neck region until head flesh is consumed (Table 3.2). A similar circumstance was observed on the two giraffe and one elephant carcass discovered. The consumption of cervical flesh after head flesh on small adult carcasses (Table 3.2) is caused by the tendency of lions, cheetah and vultures to peel a protective layer of body skin over the neck while feeding from the ribcage and shoulder. Hyenas on smaller carcasses, however, owing to

Table 3.2: Consumption sequences for three sizes of adult carcasses. The rank of parts in the G.C.S. is also shown for comparison. Note the late consumption of cervical flesh in small adults and large adults (a), the early consumption of phalanges from large adults (b), and the consumption of nasal pulp before all marrow in small adults. See text for details.

Carcass Unit	G.C.S.	RANK IN CONSUMPTION SEQUENCE small adults (n=24)	med-sized adults (n=168)	large adults (n=8)
lower viscera	1	3	1	1
upper viscera	4	6	4	3
HINDQUARTER FLESH				
pelvis	2	1	2	2
lumbar	5	4	5	5
femur	3	2	3	4
tibia	9	5	10	6.5
FOREQUARTER FLESH				
ribcage	6	8	6	8
scapula	8	7	8	9
humerus	7	9	7	6.5
radius-ulna	11	10	11	11
cervical	10	14[a]	9	18.5[a]
HEAD FLESH				
tongue	12	11.5	12	11
mandible	13	11.5	13	11
maxilla	14	13	14	18.5
HINDLIMB MARROW				
femur	15	16	15	18.5
tibia	16	17	16	18.5
metatarsal	18	23	18	18.5
phalange	17	24	19	13[b]
FORELIMB MARROW				
humerus	19	18	17	15
radius	21	20.5	20	18.5
metacarpal	22	20.5	21	18.5
phalanges	20	20.5	22	14[b]
HEAD CONTENTS				
nasal pulp	23	15[c]	23	23
mandible pulp	25	20.5	24	24.5
brain	24	25.5	25	24.5
frontal pulp	26	25.5	26	24.5

Table 3.3: Consumption sequences for three ages of wildebeest and zebra. The rank of parts in the G.C.S. is shown for comparison. Note the early consumption relative to marrow bones of the brain and frontal pulp in neonates (a) and to a lesser extent in juveniles (b). See text for details and other deviations from the G.C.S.

Carcass Unit	G.C.S.	RANK IN CONSUMPTION SEQUENCE neonates (n=20)	juveniles (n=25)	adults (n=168)
lower viscera	1	3	1	1
upper viscera	4	5	5	4
HINDQUARTER FLESH				
pelvis	2	3	3	2
lumbar	5	3	5	5
femur	3	1	2	3
tibia	9	6	10	10
FOREQUARTER FLESH				
ribcage	6	7	5	6
scapula	8	8.5	7.5	8
humerus	7	8.5	7.5	7
radius-ulna	11	16	11	11
cervical	10	10	9	9
HEAD FLESH				
tongue	12	12	13	12
mandible	13	12	12	13
maxilla	14	12	14	14
HINDLIMB MARROW				
femur	15	21	17	15
tibia	16	19.5	16	16
metatarsal	18	19.5	19	18
phalanges	17	17	20	19
FORELIMB MARROW				
humerus	19	23	18	17
radius	21	24.5	22	20
metacarpal	22	24.5	23.5	21
phalanges	20	22	23.5	22
HEAD CONTENTS				
nasal pulp	23	18	15	23
mandible pulp	25	26	26	24
brain	24	14.5[a]	21[b]	25
frontal pulp	26	14.5[a]	25	26

Table 3.4: Consumption sequence for medium-sized adults by individual carnivore taxa. The rank of parts in the G.C.S. is shown for comparison. Consumption by vultures stops at the last flesh unit (radius-ulna), while the only within-bone tissues eaten by lions on this size class are nasal pulp and occasionally one extremity of femur marrow. Note the early consumption of hindlimb marrow before all head flesh by hyenas (a), and the early consumption of head flesh relative to most forequarter flesh by vultures (b).

Carcass Unit	G.C.S.	RANK IN CONSUMPTION SEQUENCE hyenas only (n=43)	lions only (n=38)	vultures only (n=10)
lower viscera	1	1	1	2
upper viscera	4	2	3	8
HINDQUARTER FLESH				
pelvis	2	3	2	1
lumbar	5	5	5	4
femur	3	4	4	3
tibia	9	7	10	11
FOREQUARTER FLESH				
ribcage	6	6	6	5
scapula	8	9	8	12.5
humerus	7	8	7	12.5
radius-ulna	11	12	11	14
cervical	10	10	9	10
HEAD FLESH				
tongue	12	16	13	7[b]
mandible	13	17	12	6[b]
maxilla	14	22	14	9[b]
HINDLIMB MARROW				
femur	15	11[a]	16	--
tibia	16	13[a]	--	--
metatarsal	18	14[a]	--	--
phalanges	17	15[a]	--	--
FORELIMB MARROW				
humerus	19	18	--	--
radius	21	19	--	--
metacarpal	22	20	--	--
phalanges	20	21	--	--
HEAD CONTENTS				
nasal pulp	23	24	15	--
mandible pulp	25	23	--	--
brain	24	25	--	--
frontal pulp	26	26	--	--

their powerful tearing jaws, were observed to conform to the back to front pattern seen in the G.C.S.

All specific sequences save one conform to the G.C.S. with respect to the consumption of marrow after all flesh. The exception is the hyena, which is shown to consume hindlimb marrow before head flesh in medium-sized adults (Table 3.4). On smaller adult carcasses, hyenas were observed to consume hindlimb marrow even earlier, before most forequarter flesh. The early consumption of hindlimb marrow by hyenas is partly an artifact of the tendency of individual hyenas to remove recently defleshed hindlimbs from the highly competitive feeding that characterizes hyenas at a carcass (see also Kruuk 1972:125). Hence, although hyenas do not actually finish consuming hindlimb marrow until later than indicated in Table 3.4, the tissues are nonetheless removed from the pool of potentially scavengeable parts at a stage earlier than predicted by the G.C.S. Hyenas consume forelimb marrow at a stage more in keeping with the G.C.S., leaving until later among flesh parts only maxillary flesh. The conformity of the forelimb marrow parts results from the need for these to be mostly defleshed before removal of the marrow bones, which typically occurs late in the consumption of all flesh parts.

Lions on small adults, small subadults, and occasionally medium-sized juveniles are the only other measurable consumers of marrow bones in the Serengeti. [Wild dogs are certainly capable of exposing marrow cavities of animals at least this large, but on the one occasion they were observed to feed, the carcass was abandoned at an early stage in favor of joining another hunt.] Lions were never observed to gain access completely to the marrow cavities of any bone of medium-sized adult or larger carcasses. Lions frequently do gnaw the epiphyses of especially the humerus and femur of large carcasses, but can expose at most the very top or bottom of the marrow cavity. The lower rank of the femur compared to other marrow bones on medium-sized adult carcasses eaten by lions (Table 3.4) results from several occasions where the distal epiphysis of the femur had been heavily gnawed, and a small portion of the marrow removed in the process. Cheetah and vultures were never observed to crack marrow bones of even the smallest carcass.

In most specific cases, all hindlimb marrow is consumed before any forelimb marrow, as in the G.C.S. This is certainly true for hyenas, the consumer responsible by far for most marrow consumption, on medium-sized adults (Table 3.4) and other carcasses. Exceptions, however, do occur. On medium-sized juveniles (Table 3.3) and medium-sized adults (Tables 3.2 and 3.3), humerus marrow is consumed before metatarsal and phalange marrow of the hindlimb. The reasons for this are uncertain, but the pattern is seen to a greater extent for small adults (Table

3.2), where all forelimb marrow is consumed before the marrow of the metatarsus and phalanges of the hindlimb. This pattern was observed on several occasions of lions consuming Thompson's gazelle adults. The preferred tissue of lions is clearly flesh, and consumption of marrow bones on small carcasses, or the gnawing of their epiphyses on larger carcasses, appears to be an activity carried out as an afterthought to the meal just consumed. In these terms, the sequence of marrow part consumption by lions might be expected to follow a more random pattern than that practiced by the bone-crunching-specialist hyenas. The final exception is seen on large adult carcasses (Table 3.2), where marrow from both the hindlimb and forelimb phalanges is consumed before any other marrow. A similar pattern was observed on the two giraffe carcasses discovered, and relates to the manageability of phalanges compared to the other, exceedingly large and difficult to crack marrow bones of large carcasses.

Head contents are generally low yielding and difficult to extract carcass parts eaten last, as a group. This is invariably the case for mandibular pulp, and in all but one case frontal sinus pulp. Nasal pulp is the most readily exposed head content, being protected only by skin and laminar bone. It is consumed before all or most marrow in small adult (Table 3.2) and all subadult (Table 3.3) carcasses. This pattern results from the tendency of lions to consume most nasal pulp early in the consumption sequence on all kills, as documented here for medium-sized adults (Table 3.4). Aside from nasal pulp, lions were never observed to consume head contents of medium-sized adult and larger carcasses. Vultures (Table 3.4) and cheetah were never observed to consume any head content on any size carcass.

The brain is the other head content to show deviations from the G.C.S. In all adults (Table 3.2) and for all consumers (Table 3.4), the brain is eaten after marrow bones, as predicted by the G.C.S. Adult brains are securely encased in thick cranial bones, and in many bovids protected additionally by strong frontal bosses. When eaten, the brain of adult carcasses is typically exposed by gnawing a circular hole in the occiput. The cranium of subadults, and particularly neonates, is thin and unprotected by bosses. The brain is an easily exposed tissue, and relative to unfatty limb bones of subadults, is nutritionally more rewarding. Accordingly, the brain is consumed before and frequently to the exclusion of all marrow bones in neonates (Table 3.3) by lions and hyenas, by crushing the superior cranium and in the process, consuming the frontal sinus region. The variable fat content of juvenile marrow bones is also apparently responsible for the earlier consumption of the brain compared to lower forelimb marrow bones. Unlike neonates, however, the brain of medium-sized juveniles is exposed by lions and more frequently hyenas in the fashion typical for adults, as reflected by the later consumption of frontal pulp (Table 3.3).

Indexing Carcass Completeness

The important deviations from the regional carcass consumption sequence of the G.C.S. described above require carcass completeness to be expressed using several values. The consumption sequence of the three flesh regions was shown to be consistent for most carcass and consumer categories. Upper viscera, however, tend to be eaten later than most hindquarter flesh. An individual completeness value should therefore be used to describe the amount of viscera remaining on a carcass, while a separate and single value can accurately characterize the amount and distribution of all remaining flesh. Marrow and anterior body flesh were shown to be eaten in differing orders depending on consumer type. Likewise, head contents, particularly the brain and nasal pulp, were shown to be consumed at various stages relative to marrow bones depending on carcass age. Separate completeness values are therefore required for marrow bones and head contents. Within any one of the four tissue types (viscera, flesh, marrow, head contents), a particular completeness value will correspond to a fairly predictable series of constituent parts, depending on the consumer and carcass type being considered, and according to the specific consumption sequences presented in tables 3.2 through 3.4.

An index is used to characterize the completeness of each of the four carcass components. To simplify analysis, an index is used rather than estimates of the absolute weights of remaining tissue.

The index was constructed for each carcass in the following way: whole number values of 0 - 4 were assigned to each part as in constructing consumption sequences. The sum of the values of all parts within each of the four carcass components was obtained. This sum was then divided by the value of the component if complete. The quotient obtained is the percentage indexed completeness value for the component. Using viscera as a simple example, a carcass whose lower viscera have been completely consumed (completeness value of 0), but which remains with approximately half of the upper viscera (value of 2) has a total completeness of 2. Relative to the whole visceral completeness of 8, the percentage indexed completeness value for viscera of this carcass is 25%. A similar procedure was followed for other tissue components of the carcass. Unlike the consumption sequence, however, both members of paired, appendicular parts were counted, yielding 18 flesh parts with a total possible completeness of 72, 16 marrow parts with a total completeness of 64, and 4 head content parts with a total completeness of 16. By summing the completeness values of all four components (total possible of 160 for 40 carcass parts), an indexed completeness value was obtained for the whole carcass.

The index as constructed ignores drastic differences in the yield of individual carcass parts. Because the G.C.S. proceeds from generally high yielding to low yielding parts, (see Blumenschine and Caro, in press) this method overestimates the actual completeness of thoroughly eaten carcasses relative to more complete ones. However, given the broad applicability of the G.C.S. and a knowledge of the important deviations from it, a particular indexed completeness value will correspond to a fairly specific set of parts. For instance, an indexed completeness value for flesh of 45% denotes remaining flesh to be represented mostly by the eight flesh units eaten last. For virtually all consumer and carcass types, the eight units include the three head parts, cervical flesh, and flesh from both tibiae and radius-ulnae. Similarly, a total carcass indexed completeness value of 30% denotes remaining edible tissues to be restricted mostly to the last twelve parts consumed. These include, through reference to the consumption sequences, head contents and marrow mostly of the forelimb, with the brain being included if juveniles and adults, but not neonates, are being considered. The indexed completeness values therefore provide a fairly accurate and simply derived means of expressing the amount <u>and</u> distribution of edible tissues remaining on a carcass.

The completeness values refer to the state of a carcass upon abandonment by consumers. Completeness on abandonment is used instead of completeness on discovery because the former event provides the only standard reference point in the resource life of a series of carcasses. Having a standard reference point is essential for comparing the scavenging opportunities provided by carcasses in different ecological and physical contexts. Carcass abandonment is also the point at which any scavenger, regardless of its rank on the competitive hierarchy, at least potentially has access to the carcass. Since carcasses are often abandoned and taken over by other consumer taxa several times before its resources are exhausted, several useful reference points of scavenging opportunity exist for most carcasses.

CARCASS PERSISTENCE

The duration over which a carcass persists as a source of scavengeable food provides another measure of the scale of scavenging opportunity. If continuously consumed from the time of death, the persistence of a carcass would be directly proportional to its completeness at any point in its resource life. Many carcasses, however, are abandoned by initial and subsequent consumers in a partially consumed state, and can remain unattended by scavengers for varying amounts of time. The longer this interval, the more likely a carcass is to provide a feeding opportunity to a particular scavenger.

Carcass persistence is measured in hours from the time of death until all edible parts have been consumed. The death of only 4 of 268 carcasses was witnessed, requiring estimates to be made of the interval between death and discovery of the carcass. The day of death for most carcasses could be determined easily because most parts of a particular study area were monitored on a daily basis. As well, information on carcass size and completeness on discovery, and the number and types of consumers present ensured that most carcasses were less than one day old. Occasionally, a carcass was discovered two or more days after death, as indicated by the dryness of remaining tissues, the presence of maggots and the general degree of tissue decomposition. Deaths of the previous night were assumed to have occurred at midnight, corresponding approximately to the earlier of two peaks of nocturnal predator activity documented for lions and hyenas in the Serengeti and Ngorongoro (Kruuk 1972, Schaller 1972).

The resource life of a carcass was deemed exhausted when all edible parts were observed to have been consumed, or when the marginal remains were observed to be transported from the death site, usually by hyenas carrying scraps to a den. A partially consumed carcass eaten in full or transported away from the death site at night was assumed to have persisted until midnight. In many cases, I was unable for logistical reasons to observe or estimate accurately the end of a carcass's resource life, and these are necessarily excluded from analyses of carcass persistence.

CHAPTER 4

SCAVENGING OPPORTUNITIES IN THE SERENGETI AND NGORONGORO

A number of factors have been found to effect the scale of scavenging opportunity in the modern Serengeti and Ngorongoro ecosystems. These factors can be grouped within certain physical and ecological contexts among which scavenging success varies. The contexts of scavenging are in this way viewed as independent variables that condition carcass persistence and completeness on abandonment by initial and subsequent consumers.

Two sets of contexts have been found to have, in their combined effect, a major influence on the scale of scavenging opportunities. These are carcass size in relation to the size and number of initial consumers, and the degree of competition for carcasses. Each set of contexts is based on sound principles of uniformitarianism such that their influence on modern scavenging opportunity should be similar in kind to that in the past.

Carcass size in relation to the number and size of initial consumers defines a fundamental physical context within which the resulting feeding opportunity for a scavenger should vary in a predictable manner. A carcass of a given size provides a certain quantity of food and imposes certain mechanical constraints upon consumption. The extent to which a carcass is consumed is directly dependent on individual and group satiation levels, and the ability of consumers to overcome mechanical constraints to the consumption of carcass parts. These constraints will be greatest for edible tissues within bones, especially on larger carcasses. Larger carcasses are also likely to be less completely eaten before consumers are satiated and thereby to survive as food resources through multiple feeding episodes. Carcasses consumed by weak-jawed carnivores, too, are more likely to persist as food resources through one or several feeding episodes, and hence provide a better feeding opportunity for scavengers. Among the consumers regularly observed, hyenas are least constrained by physical obstructions to consumption owing to their powerful jaws. Weaker-jawed felids have only limited access to tissues within bones, and vultures have the most restricted access to unexposed edible tissues. The largely physical basis of these expectations, including jaw strength relative to bone strength, and carnivore gut size relative to the weight of edible carcass tissues, indicates modern observations on size determinants of scavenging opportunity to be an accurate informant of those carcass sizes most frequently available in the past.

The intensity of competition for carcasses defines a basic ecological context within which scavenging success should vary in a predictable manner. Greater competition for carcasses should result in lower average per carcass completeness when encountered by a scavenger, and generally lower encounter rates stemming from shorter average carcass persistence. The intensity of competition for carcasses can be approximated by local consumer to carcass ratios. Although the relative numbers of consumers and carcasses in an area proved impractical to determine, they are nonetheless expected to vary directly with ecosystem type, season, and habitat type. Within each of the three contexts, the completeness by which carcasses are consumed and the duration carcasses persist as food resources will be shown to vary in a predictable manner according to the intensity of competition. By using competition as the fundamental basis for identifying some of the ecological contexts within which scavenging success varies, determinations of ecologically-based variations in prehistoric scavenging success are simplified.

CARCASS SIZE, CONSUMER TYPE AND SCAVENGING OPPORTUNITY

Carcass Size and Persistence

The average persistence of carcasses of different size is shown in Table 4.1. Persistence, again, is measured from the estimated time of death until the carcass no longer provides a feeding opportunity. For all medium-sized adult and smaller carcasses, the latter point corresponds exclusively to the complete consumption of all edible tissues. For adult buffalo and larger carcasses, scavenging opportunities ceased to exist when decomposition of remaining tissues inhibited vertebrate carnivores from deriving further nourishment from the carcass. Since I was not always able to observe the end of a carcass's resource life, the data in Table 4.1 are a subsample of the total carcass sample. Also contained in Table 4.1 is the weighted average of the estimated live weights of the carcasses comprising each size/age class.

The average persistence of carcasses increases from one hour in small subadults to 272 hours in adult giraffe and elephant, well after the onset of decomposition. Small adult and medium-sized juvenile carcasses persist for less than seven hours. This is at least three times shorter than medium-sized adult persistence, which on average is approximately one day. Adult buffalo persisted for an average of 100 hours, the approximate time decomposing microorganisms and maggots require to become well established in a carcass.

Table 4.1: Average persistence of carcasses as food sources in each size/age group. Only those carcasses for which the whole resource life was monitored are included. Also shown are the estimated live weights of the individual taxa comprising each group. Weights for adult giraffe, elephant and zebra are from Dorst and Dandelot (1969). Weights for the remaining adults are from the male/female averages of Ledger (1968). Juvenile weights as estimated as half that of adults, while neonate weights are estimated as one-sixth of adult weight. For persistence, all paired differences are significant at least to the .05 level (Mann-Whitney U-test, direction predicted) save small subadults and small adults, and small adults and medium-sized subadults. See Figure 4.1 for the relationship between average size and persistence.

Carcass Size Group	Total n	Composition (n)	Live Weight (kg)	Weighted Mean of Group Wt.	Persistence in hours (mean/S.D.)
Small		Thompson's gazelle subadults (3)	3.1		1.0
subadults	6	Grant's gazelle neonate (1)	8.5	13.0	(0.6)
		Wildebeest neonates (2)	30		
Small	5	Thompson's gazelle (3)	18.6	31.4	4.4
adults		Grant's gazelle (2)	50.7		(4.1)
Medium-sized		Wildebeest (8)	90	100.0	6.5
juveniles	10	Zebra (2)	140		(2.9)
Medium-sized		Wildebeest (60)	181.7	196.0	20.8
adults	71	Zebra (11)	272.7		(26.3)
large adults	7	Buffalo (7)	753	753.0	100.1
					(61.3)
Very large	3	Giraffe (2)	1182	2303.0	272.0
adults		Elephant (1)	4545		(63.0)

Figure 4.1: Carcass size and persistence. Mean values for each size group are shown. For the regression line drawn a = -1.2, b = 1.1, r = .99.

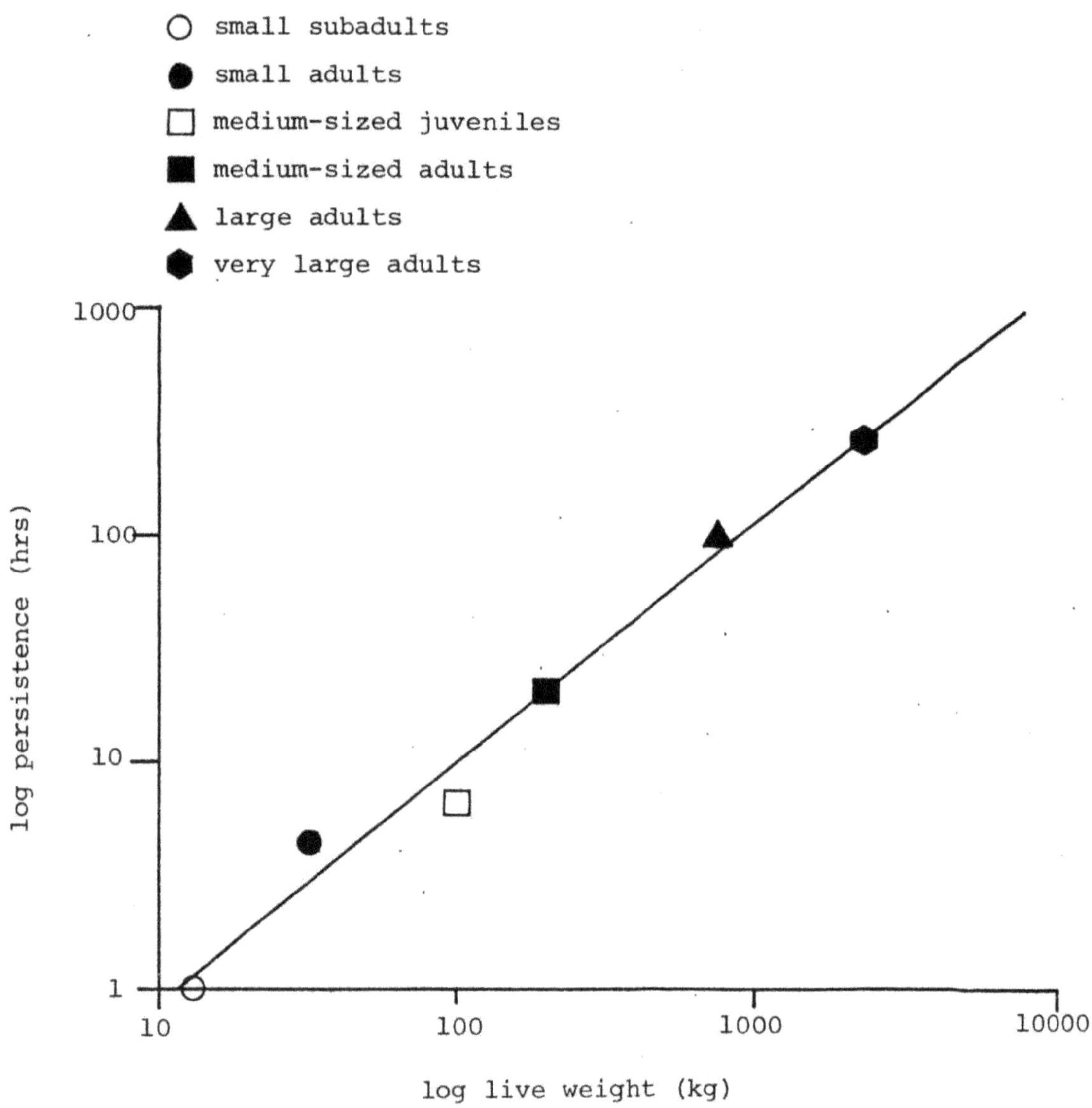

Figure 4.1 shows the relationship between carcass size and persistence to be nearly linear and highly correlated on a double logarithmic scale. Logarithmic transformations of size and persistence are appropriate because rates of consumption decrease exponentially as progressively lower yielding and harder to extract tissues are consumed, and because the allometry of weights of anatomical parts are commonly best described by logarithmic functions.

The strong correlation between average carcass size and persistence obscures many important variations in persistence within any one size class (see standard deviations in Table 4.1). For the moment I am particularly concerned with illustrating those variations caused by the type of consumer responsible for the initial stages of carcass consumption. For example, the average persistence of medium-sized adults from the Serengeti consumed initially by hyenas is 8.2 hours (n = 18, SD = 6.3). Occasionally I observed hyenas finishing an adult wildebeest or zebra in less than an hour. On the other hand, medium-sized adult carcasses initially consumed by lions and only subsequently finished by hyenas persisted for an average of 34.5 hours (n = 30, SD = 33.2). The significantly longer persistence of lion kills (Mann-Whitney U-test, $p < .00001$) is partly due to slower rates of consumption and smaller feeding group sizes than seen among hyenas, but primarily results from the inability of lions to extract marrow and most head contents. These parts persist on average for more than a day until they are located and scavenged by hyenas. The manner by which hyenas reduce carcass persistence, and scavenging opportunities in general, will be fully discussed in the section on competition. Other factors effecting carcass persistence will be also be discussed. Suffice it to say for now that carcass size is a fundamental determinant of carcass persistence, but one whose effect is greatly modified by the ecological setting of the carcass.

Carcass Size and Completeness

Most carcasses in the Serengeti and Ngorongoro are eventually completely consumed by vertebrate carnivores. Of the 232 carcasses for which such data are available, only 10 (4.3%) were not completely consumed, being allowed to decompose; most of these were large carcasses (Table 4.2). No small subadult or small adult carcasses were observed to decompose, being completely consumed well before maggots and microorganisms became established. Only 4% of medium-sized juveniles and 2% of medium-sized adults were not completely consumed by vertebrate carnivores. On the other hand, 43% of the adult buffalo and all of the very large adult carcasses found remained with edible tissues that eventually decomposed.

Table 4.2: Percentage of carcasses of each size class observed to decompose. Carcasses are excluded for which persistence was known to be unusually long, but which could not be confirmed to have decomposed or to have been completely consumed by vertebrate carnivores.

Carcass Size Group	Total n	Carcasses Decomposing n	%
Small subadults	24	0	0.0
Small adults	16	0	0.0
Medium-sized juveniles	26	1	3.8
Medium-sized adults	156	3	1.9
Large adults	7	3	42.9
Very large adults	3	3	100.0

The rarity of incomplete carcass consumption does not denote minimal scavenging opportunities or their restriction to only the largest carcasses. Many carcasses survive two or more abandonments by lions and/or hyenas, the frequency of which is partly size-dependent (Table 4.3). Less than 25% of medium-sized adult and smaller carcasses survive two or more abandonments, while all large adult and larger carcasses do. These data indicate that a smaller proportion of parts of larger carcasses will be consumed in any one feeding bout. More importantly, they show that some medium-sized adult and smaller carcasses survive initial and secondary consumption, and are therefore available to scavengers.

A more detailed understanding of the size-dependency of scavenging opportunity is provided by data on carcass completeness on abandonment by initial consumers. Comparisons are necessarily restricted to carcasses fed on by the same initial consumer taxon. Table 4.4a-d and Figure 4.2a-d show the mean percentage indexed completeness of variously sized carcasses fed on initially and only by lions, hyenas in a single feeding bout, vultures, and cheetah.

Lions were observed to feed from the largest size range of carcasses, and to do so with a completeness dependent on carcass size. The mean indexed completeness of the total carcass on abandonment by lions increases from approximately 12% for small subadults to almost 70% for large adults (Table 4.4a, Figure 4.2a). Although lions were known to have fed on one of the very large carcasses in my sample, their effect to the exclusion of other consumers was not observed.

Virtually all viscera are consumed by lions regardless of carcass size. Lower viscera, surrounded by deposits of mesenteric fat, are almost always completely eaten first. Upper viscera are consumed somewhat later (refer to Table 3.4), and occasionally, as reflected by the small amount of visceral remains of medium-sized adults in Table 4.4a, a single lung is not consumed.

The remaining carcass components (flesh, marrow, head contents) show similar trends toward increasing completeness on abandonment with increasing carcass size (Table 4.4a, Figure 4.2a). Negligible amounts of flesh are abandoned by lions on the three smallest carcass size classes. Medium-sized adults are typically abandoned with flesh remaining on the head and the distal radius-ulnae and distal tibiae. Buffalo adult carcasses usually survive lion feeding bouts with flesh on much of the anterior half of the body. This is the case even when lions remain with the carcass for several days to obtain several meals from it.

As stated in Chapter 3, marrow bones and head contents (with the exception of nasal pulp) were never observed to

Table 4.3: Size-specific proportion of carcasses surviving as food resources after two or more abandonments by lions and/or hyenas.

Carcass Size Group	Total n	Carcasses Surviving Multiple Abandonments n	%
Small subadults	24	3	12.5
Small adults	16	3	18.8
Medium-sized juveniles	26	5	19.2
Medium-sized adults	156	38	24.4
Large adults	7	7	100.0
Very large adults	2	2	100.0

Table 4.4: Completeness by which single consumer taxa abandon carcasses of different size and age. Note that an absence of records for a particular carcass size class denotes only that the effect of the single consumer taxon alone was not observed on the size class.

Carcass Size Group	n	Viscera	Flesh	Marrow	Head Contents	Total
		PERCENTAGE INDEXED COMPLETENESS: MEAN ± S.D.				
A) LIONS ONLY						
Small subadults	2	0	1.4 ± 1.4	25.0 ± 25.0	12.5 ± 12.5	11.9 ± 11.9
Small adults	1	0	1.4	21.9	75.0	16.9
Medium-sized juveniles	5	0	0.6 ± 0.7	82.5 ± 35.0	62.5 ± 27.1	38.6 ± 16.2
Medium-sized adults	9	2.6 ± 8.7	12.1 ± 17.0	99.8 ± 0.7	95.1 ± 6.3	55.5 ± 8.2
Large adults	1	0	4.3	100.0	100.0	69.4
B) VULTURES ONLY						
Medium-sized juveniles	3	0	1.9 ± 2.6	100.0	100.0	50.8 ± 1.2
Medium-sized adults	2	0	5.6 ± 0.0	100.0	100.0	52.5 ± 0.0
Very large adults	2	43.8 ± 43.8	70.2 ± 28.5	100.0	100.0	83.8 ± 15.0
C) HYENAS ONLY						
Small subadults	5	5.0 ± 10.0	14.4 ± 28.9	14.7 ± 29.4	11.3 ± 19.5	13.7 ± 27.2
Medium-sized juveniles	8	0	0	9.2 ± 16.3	3.9 ± 7.0	4.1 ± 6.3
Medium-sized adults	36	1.4 ± 6.4	2.0 ± 9.6	8.3 ± 20.5	24.7 ± 38.2	7.2 ± 15.2
D) CHEETAH ONLY						
Small adults	4	3.1 ± 5.4	20.2 ± 22.4	100.0	87.5 ± 12.5	58.5 ± 10.7

Figure 4.2: Size-specific completeness of carcass consumption by different initial consumers. Accompanying data appear in Table 4.4a-d. V = viscera, F = flesh, M = marrow, HC = head contents, T = total carcass.

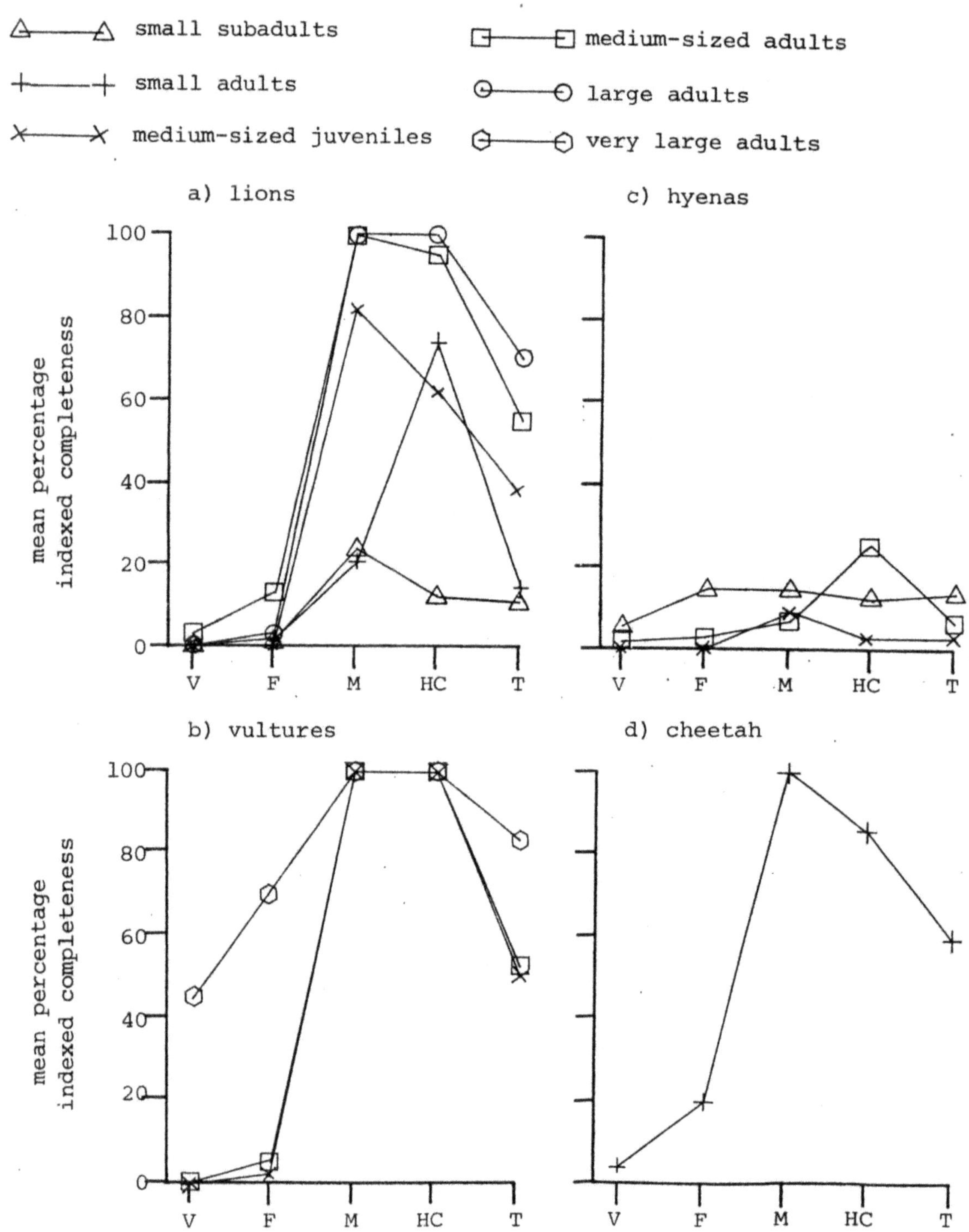

be consumed by lions on medium-sized adult and larger carcasses (Table 4.4a, Figure 4.2a). Marrow bones are only infrequently consumed from medium-sized juveniles, being restricted usually to upper limb bones. Lions occasionally expose the brain of juvenile wildebeest and zebra, but aside from nasal pulp, do not consume other head contents of this size class. Data in Table 4.4a show lions to consume marrow bones to a similar extent in small subadult and small adult carcasses. This apparently violates the expectation of relatively incomplete limb bone consumption in subadults owing to low fat content (see Chapter 3). Compared to head contents, however, the consumption of marrow bones is much less complete in small subadults than in small adults. Head contents of medium-sized juveniles are also, on average, more completely consumed by lions than marrow bones.

The reversed sequence of marrow and head contents consumption in adult and subadult carcasses will be seen to apply to hyena consumption patterns as well. The pattern can be ascribed to both nutritional and physical causes: marrow bones of adults are preferred for their high fat, hence high energy content over lower yielding and harder to extract head contents. The ranking of carcass components is switched in subadults and particularly neonates owing both to the low to nil fat content of marrow bones and the relative ease by which the brain is exposed.

Despite important differences in the components consumed, there remains a strong size dependency in the completeness of total carcass consumption by lions. The amount of viscera and flesh consumed by lions appears to depend on satiation levels, which, assuming constant feeding group size, are reached at an earlier stage of the consumption sequence in larger carcasses. The amount of marrow and head contents consumed by lions is also a function of satiation, but additionally of their extremely limited ability to expose these tissues. Hence, the expected preference for marrow bones over head contents seen for lions on small adults is not replicated for the medium-sized adult category simply because of the lion's inability to crack marrow bones of adult wildebeest and zebra. That lions are necessarily limited to flesh-eating on medium-sized adult and larger carcasses is largely responsible for the high completeness by which within-bone tissues are abandoned to scavengers.

Vultures also seem to be constrained by the mechanics of larger carcasses and as a result produce the expected relationship between carcass size and completeness on abandonment (Table 4.4b, Figure 4.2b). Vultures consumed all viscera on the few naturally dead medium-sized juveniles and adults on which they were observed as initial consumers, but were able to gain access to only approximately half of the visceral parts of the two very large adult carcasses. The remaining viscera were all inaccessible upper viscera. A similar trend is seen with flesh. Very little flesh, restricted to the lower limbs, is abandoned on medium-sized juveniles and adults; almost three-quarters of the flesh parts of very large carcasses, however, cannot be exposed by vultures feeding alone. Regardless of carcass size, vultures cannot expose tissue within bones. Vultures are therefore effective flesh consumers on all but the largest carcasses, but will always abandon marrow and head contents to other scavengers.

The completeness by which hyenas as initial consumers abandon carcasses of different size is shown in Table 4.4c and Figure 4.2c. Although hyenas were involved with the consumption of carcasses from all size classes, I only observed them as initial consumers to abandon carcasses from the three size groups shown.

Unlike lions and vultures, hyenas do not show a consistent size-dependent completeness of carcass consumption. The expected relationship holds to a slight degree between medium-sized juveniles and medium-sized adults. Hence, hyenas were always observed to consume all viscera and flesh of juvenile wildebeest and zebra, and to consume virtually all head contents as well. Hyenas often leave small portions of viscera and flesh on adult wildebeest and zebra, and substantial amounts of head contents, including the brain in 9 of the 36 cases. Small subadults, however, are abandoned by hyenas with a higher average completeness than larger carcasses. The departure from expectation is partly a reflection of the higher completeness by which nutrient-poor marrow bones of young animals are abandoned (which in both subadult categories is higher than head content completeness, as expected). Small subadults, however, are also abandoned with more flesh than the two larger carcasses classes.

The lack of a size-dependent completeness of consumption by hyenas is a reflection of their unconstrained consumption of virtually all carcass parts of even the largest carcasses (Coe 1980, personal observations). The extent to which hyenas consume a carcass, rather, should depend exclusively on satiation levels and the number of hyenas feeding. Table 4.5 shows the average feeding group size of hyenas on their kills. An average of 2.2 hyenas fed on each small subadult carcass, while the values for medium-sized juveniles and adults was 12.1 and 10.9, respectively. The sizes of hyena feeding groups are related to the types of prey available (Kruuk 1972). Juvenile and adult wildebeest and zebra were usually observed to be fed on by a whole clan, as capture of these prey usually requires a cooperative group effort. Foraging for easily captured very young prey, whether cached gazelle fawns or neonate wildebeest, is an individual pursuit, and the carcass is quickly consumed by a single individual or a pair of hyenas.

Table 4.5: Average number of hyenas or lions feeding as initial consumers on carcasses of different size and age. Number of carcasses for which feeding group sizes could be determined are lower than the actual number of carcasses known to have been consumed initially by each of the two consumers. The number of lions feeding excludes very young cubs; half-grown cubs each count as one half of an adult. All hyenas are counted equally, as very young individuals do not appear at carcasses. Compare average group sizes for each carcass size class with the completeness of consumption shown in Table 4.4a for lions, and Table 4.4c for hyenas.

Carcass Size Group	HYENAS ONLY n	HYENAS ONLY mean #/ carcass	LIONS ONLY n	LIONS ONLY mean #/ carcass
Small subadults	5	2.2	2	4.0
Small adults	0	---	1	1.0
Medium-sized juveniles	7	12.1	5	6.8
Medium-sized adults	33	10.9	18	6.7
Large adults	0	---	1	16.0
Very large adults	0	---	0	---

Table 4.5 also shows the size-specific average number of lions feeding per carcass. Lion feeding group sizes are directly proportional to carcass size, but inversely related to the completeness of consumption. These data support the suggestion that mechanical constraints imposed by the carcass are a major determinant of the completeness by which lions consume carcasses, with satiation playing a smaller role.

The completeness by which cheetah consume carcasses is shown in Table 4.4d and Figure 4.2d. Although I was able to observe the effect of solely cheetah consumption on only one size class, small adults, these data are presented to illustrate the importance of relative consumer to carcass size on the completeness of consumption. The reader can compare the completeness of cheetah consumption of Thompson's gazelle adults (Table 4.4d) to that by the larger lion on comparably scaled-up medium-sized adults (Table 4.4a). A general similarity in completeness of each carcass component, and among the ratios of all components fed on by the two felids, is apparent. This indicates carcass size in relation to consumer size to be a good predictor of carcass part completeness on abandonment by carnivores with similar relative extractive potentials. It also supports the suggestion that the physical principles underlying the completeness of carcasses encountered by scavengers in modern settings will apply to the past.

Summary of Carcass Size and Scavenging Opportunity

I have demonstrated carcass size to be a strong determinant of the completeness of consumption by initial consumers and hence of the scale of feeding opportunity encountered by scavengers. Larger carcasses persist longer and will more frequently survive multiple abandonments by consumers. If fed on by consumers contrained by obstructions to edible parts, a larger carcass will also be abandoned more complete. Hyenas are not so constrained, but rather consume at least medium-sized adult and smaller carcasses to an extent dependent on feeding group sizes. Although data are not available on the completeness by which hyenas alone consume adult buffalo and larger carcasses, they were observed as secondary, scavenging consumers of giraffe and elephant carcasses to abandon most marrow bones as well as substantial amounts of flesh. This is to be expected on the basis that hyena feeding group sizes and appetites are finite and lower than the huge quantities of food provided by very large carcasses.

COMPETITION AND SCAVENGING OPPORTUNITIES

Competitors and Their Effect As Scavengers

The major scavengers in the Serengeti and Ngorongoro, and in African savanna woodlands in general, are hyenas and vultures. Jackals are present as scavengers in small numbers (usually less than five) at most carcasses, but their impact on the reduction of carcass completeness was never observed to be measurable. Lions are known to be regular scavengers especially of hyena kills, usually usurping these soon after the kill is made (Schaller 1972, Kruuk 1972). However, I observed lions to scavenge only three times, all from cheetah kills of Thompson's gazelle. In light of the lion's propensity to scavenge, certainly some carcasses I discovered being fed on by lions with hyenas waiting to scavenge were actually hyena kills stolen by lions. If this were the case, the low degree of disarticulartion and bone damage that clearly distinguishes lion from hyena meals indicates lions to have scavenged the carcasses very early in the consumption sequence. Hence, all carcasses labelled "lion kills" were, if not actually lion victims, fed on primarily by lions and show the consumption pattern typical of true lion kills.

The scavenging abilities of hyenas and vultures can be evaluated by examining their effect on abandoned lion kills. Lions were responsible for the initial consumption of 44% of all carcasses for which I could identify the initial consumer. On all but the smallest carcasses, lions always abandon substantial remains to scavengers. Cheetah and vultures, too, abandon many edible parts, but their uncommonness as initial consumers (10% and 5% of the sample, respectively) renders them an infrequent contributor to scavenging opportunities. Hyenas are frequent initial consumers (38% of the total sample), but usually reduce carcass completeness to very low levels save in some circumstances to be discussed later.

The total and component part completeness of abandoned lion kills scavenged by hyenas and/or vultures are shown in Tables 4.6 through 4.9 and Figure 4.3a-d for three carcass size groups. These include medium-sized juveniles, medium-sized adults and large adults. Insufficient data are available to show the effect of scavengers on other carcass categories. For comparison, data presented previously are included on the completeness of initial consumption by lions, and of carcasses initially eaten by the scavenger being examined. It should be noted that the individual carcasses comprising the scavenged sample of each table are all not necessarily a subset of those initially consumed by lions. Some lion kills were not discovered until they had been or were being scavenged. These could nonetheless be unambiguously identified as lion kills or carcasses usurped very early by lions due to the presence of full-bellied

Table 4.6: Completeness on abandonment of medium-sized juvenile carcasses showing the effect of hyenas scavenging lion kills. Completeness as abandoned by lions only and hyenas only is shown for comparison. Data are graphed in Figure 4.3a. Note how the marrow and head content remains of lion kills are severely reduced in amount by scavenging hyenas, in a way mirroring hyenas feeding alone on their kills. For total carcass completeness, differences are significant at least to the .01 level in pairings of the hyena fed carcasses with those fed on by lions only (Mann-Whitney U-test, direction predicted).

MEDIUM-SIZED JUVENILE
PERCENTAGE INDEXED COMPLETENESS: MEAN ± S.D.

Consumer(s)	n	Viscera	Flesh	Marrow	Head Contents	Total
Lions only	5	0	0.6 ± 0.7	82.5 ± 35.0	62.5 ± 27.1	38.6 ± 16.2
Hyenas only	8	0	0	9.2 ± 16.3	3.9 ± 7.0	4.1 ± 6.3
Lions followed by hyenas once	5	0	0	9.8 ± 15.7	5.0 ± 10.0	4.3 ± 7.3

Table 4.7: Completeness on abandonment of medium-sized adult lion kills showing the effect of hyena scavenging. Completeness as abandoned by lions only and by hyenas on their kills is shown for comparison. Data are graphed in Figure 4.3b. Note the very severe reduction in yield effected by hyenas scavenging the complete marrow and head content remains of lion kills. Completeness on abandonment by hyenas scavenging is very similar to that of abandoned hyena kills. For total carcass completeness, differences in parings of hyena fed carcasses with those fed on by lions only are significant to the .001 level (Mann-Whitney U-test, direction predicted).

MEDIUM-SIZED ADULT
PERCENTAGE INDEXED COMPLETENESS: MEAN ± S.D.

Consumer(s)	n	Viscera	Flesh	Marrow	Head Contents	Total
Lions only	19	2.6 ± 8.7	12.1 ± 17.0	99.8 ± 0.7	95.1 ± 6.3	55.5 ± 8.2
Hyenas only	36	1.4 ± 6.4	2.9 ± 9.6	8.3 ± 20.5	24.7 ± 38.2	7.2 ± 15.2
Lions followed by hyenas once	25	0	0	6.6 ± 18.4	17.8 ± 32.3	4.1 ± 9.7

Table 4.8: Completeness on abandonment of medium-sized adult lion kills showing the effect of scavenging by vultures. Completeness as abandoned by lions initially, and by vultures as initial consumers is shown for comparison. Data are graphed in Figure 4.3c. Note the restriction of scavenging by vultures to the small quantities of viscera and flesh left by lions, such that all marrow bones and head contents remain untouched. No differences in any components between the three pairs of consumer groups are significant to the .05 level (Mann-Whitney U-test).

MEDIUM-SIZED ADULT
PERCENTAGE INDEXED COMPLETENESS: MEAN $\pm$ S.D.

Consumer(s)	n	Viscera	Flesh	Marrow	Head Contents	Total
Lions only	19	2.6 ± 8.7	12.1 ± 17.0	99.8 ± 0.7	95.1 ± 6.3	55.5 ± 6.3
Vultures only	2	0	5.6 ± 0.0	100.0	100.0	52.5 ± 0.0
Lions followed by vultures	17	0	6.9 ± 12.8	99.6 ± 1.5	94.1 ± 6.6	52.7 ± 6.1

Table 4.9: Completeness on abandonment of large adult lion kills scavenged by vultures and/or hyenas. Data are graphed in Figure 4.3d. Completeness as abandoned by lions alone is shown for comparison. Both vultures and hyenas effect major reductions in completeness as abandoned by lions. Vulture scavenging is restricted to flesh, while hyena scavenging emphasizes marrow bones. Due to small sample sizes, no apparent differences are significant at the .05 level (Mann-Whitney U-test).

LARGE ADULT
PERCENTAGE INDEXED COMPLETENESS: MEAN ± S.D.

Consumer(s)	n	Viscera	Flesh	Marrow	Head Contents	Total
Lions only	1	0	43.0	100.0	100.0	69.4
Lions followed by vultures	1	0	8.3	100.0	100.0	53.8
Lions followed by vultures and hyenas	5	0	15.3 ± 26.6	26.7 ± 38.9	90.0 ± 9.4	26.4 ± 28.0
Lions followed by hyenas twice	3	0	3.0 ± 2.6	0	66.7 ± 31.2	8.1 ± 4.0

Figure 4.3: Completeness of lion kills of various size showing the effect of scavenging by hyenas and/or vultures. Accompanying data appear in Tables 4.6 - 4.9. Note the severe reduction in carcass completeness resulting from hyena scavenging, and its similarity to completeness on abandonment by hyenas as initial consumers. V = viscera, F = flesh, M = marrow, HC = head contents, T = total carcass.

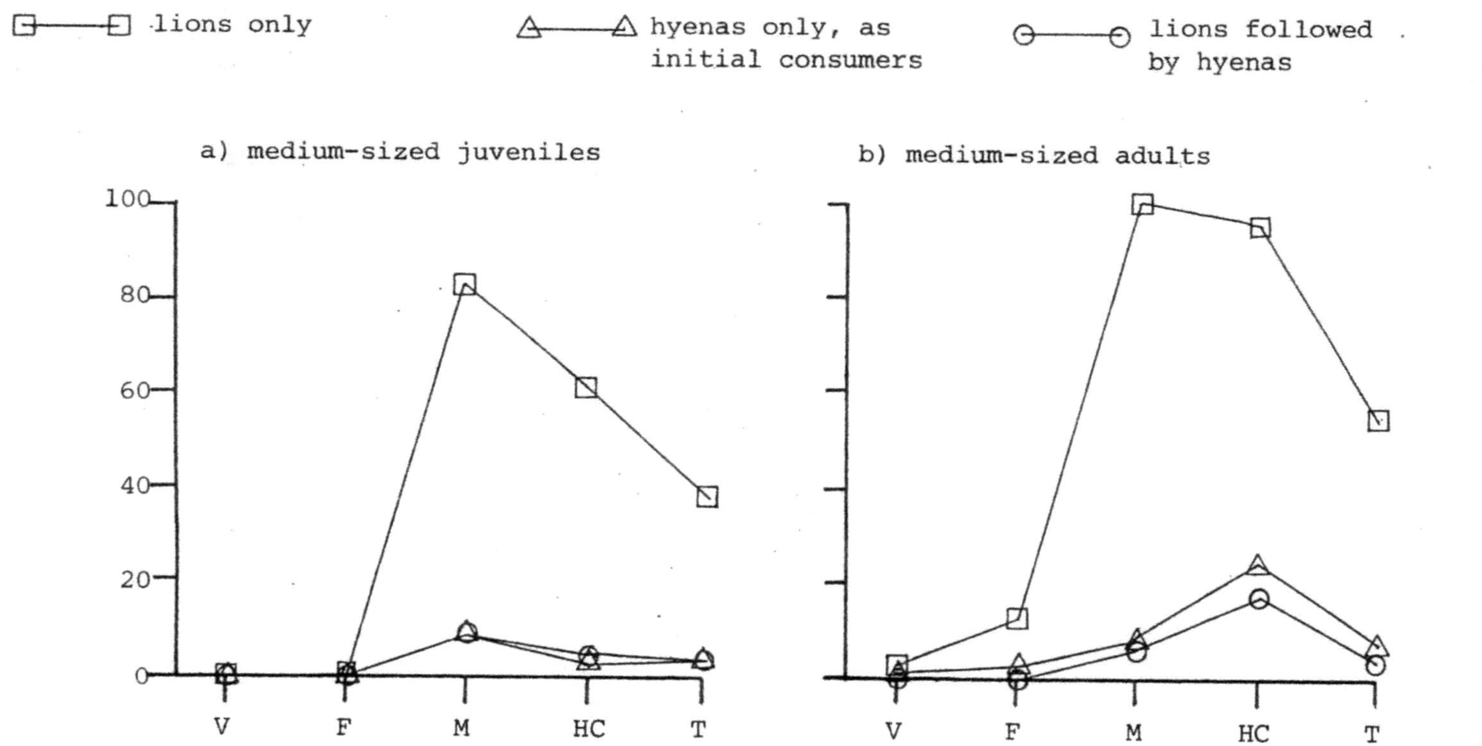

(continued)

Figure 4.3 continued

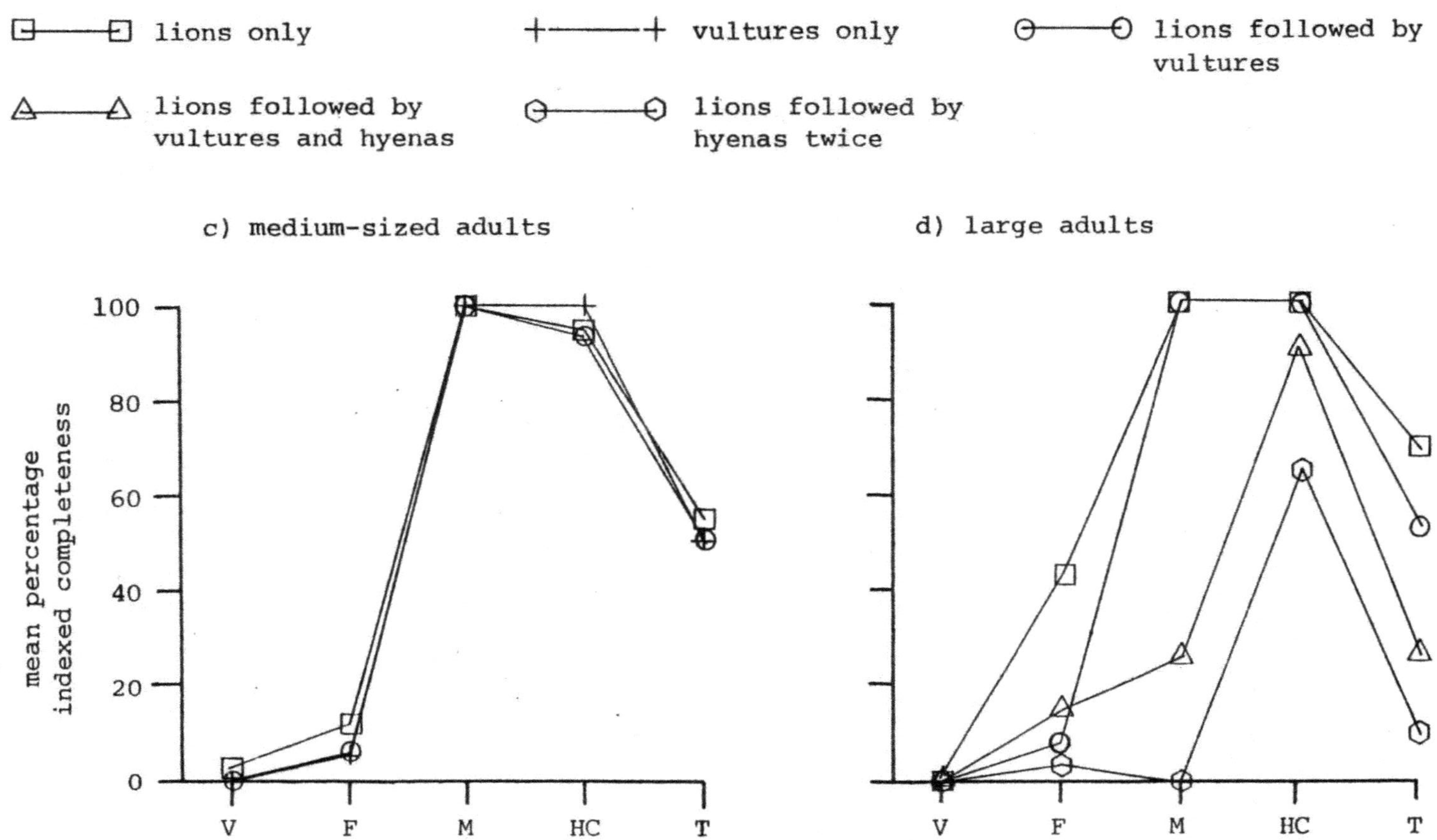

lions in the immediate vicinity, or by the degree of bone modification and disarticulation exhibited by the carcass. The data in Tables 4.6 - 4.9 therefore do not describe the exact reduction in carcass completeness by scavengers over that done by lions.

The extent to which hyenas scavenge abandoned lion kills is shown in Table 4.6 and Figure 4.3a for medium-sized juveniles, and Table 4.7 and Figure 4.3b for medium-sized adults. Hyenas usually completely scavenge the substantial remains of abandoned lion kills of both sizes. Total carcass completeness is reduced from approximately 40% and 55% as abandoned by lions, to 4% after hyena scavenging for both medium-sized juveniles and medium-sized adults, respectively. Marginal-yield head content parts, including mandibular and frontal pulp, dominate the parts of medium-sized adults ignored by scavenging hyenas; virtually all marrow and any viscera and flesh abandoned by lions are scavenged. Hyenas also completely scavenge viscera and flesh of medium-sized juveniles, and, as expected, reduce the completeness of head contents to a greater extent than marrow parts. It is interesting to note that the absolute completeness values of component carcass parts, and the relative differences among the components are very similar between the hyena-scavenged sample and the hyena-hunted sample for medium-sized adults, and virtually identical for medium-sized juveniles. Hyenas, as predators or scavengers, consume carcasses in a similar manner, and one that virtually obviates a feeding opportunity for other scavengers.

The reduction in yield of medium-sized adult lion kills by scavenging vultures is much less severe than that by hyenas (Table 4.8, Figure 4.3c). The small amounts scavenged by vultures are limited to the already small quantities of viscera and flesh abandoned by lions. Vultures consume about half of the flesh parts abandoned by lions, leaving for other scavengers flesh scattered in small pockets beneath skin of the forequarter, usually on the distal radius-ulnae. As with hyenas, the consumption patterns of vultures as initial consumers and scavengers of lion kills are very similar. Unless confronted with a complete natural death the impact of vultures on reducing scavenging opportunities from medium-sized adults is usually minimal.

The combined effect of vultures and hyenas scavenging lion kills of adult buffalo (Table 4.9, Figure 4.3d) is similar in kind to those for medium-sized adults. The reduction in total carcass yield by vultures only is greater for large adult carcasses owing to the larger amount of flesh abandoned by lions on this size class. The flesh not scavenged by vultures is that protected by the thick skin of the buffalo, including, typically, cervical flesh and distal lower limb flesh. On those lion kills of adult buffalo scavenged by hyenas and vultures, total carcass completeness is reduced from almost 70% to nearly 25% due to hyenas scavenging most marrow bones. [The completeness of flesh is higher than that after scavenging by vultures alone because different carcasses are included in the two samples.] Note again how, as in all adults, hyenas concentrate on marrow to the virtual neglect of head contents. Hyenas scavenging a second time (usually the next day) finish the remaining marrow and reduce head content completeness by almost a third, leaving mandibular pulp intact in all three cases, maxillary pulp in two, and the brain and frontal pulp in only one case.

In summary, hyenas are effective scavengers on all carcass sizes examined. As scavengers they typically abandon negligible remains to other scavengers, limiting these usually to head contents. Vultures, on the other hand, can scavenge virtually all remaining flesh, but their impact on scavenging opportunities is lower owing to the already low quantities of flesh abandoned by initial consumers. Their greatest impact would be felt on those natural deaths to which vultures have early access, but here too, their limited ability to expose tissue beneath skin renders their scavenging slow and, on large carcasses, incomplete. In analyzing the effect of competition on scavenging opportunities, it is therefore most relevant to concentrate on competition as applied by hyenas, and to identify those ecological contexts within which scavenging pressure by hyenas is variable. The series of contexts to be examined include those based on 1) ecosystems differences in predator to prey ratios, 2) seasonal differences in scavenger to carcass ratios, and 3) micro-habitat differences in the intensity of scavenging pressure applied by hyenas.

Predator to Prey Ratios and Scavenging Opportunity

Expectations and General Observations: A high ratio of predators to prey in a given area should correspond directly to high scavenger to carcass ratios. This is particularly true of Ngorongoro Crater, which supports one of the highest biomasses of mammalian predators in Africa, and a predator to prey ratio at least four times higher than that in the more typical Serengeti (see Chapter 2). High predator to prey ratios do result in high rates of predator-related mortality, but the tendency of most predators, particularly hyenas, to scavenge results in a very high degree of competition for kills. Per-herbivore predation rates in the Serengeti, on the other hand, are relatively low, and much mortality (estimated at 70% by Houston, 1979) is related to starvation or disease. Given relatively fewer hyenas and more deaths, competition for carcasses in the Serengeti should be much less intense than in Ngorongoro.

This expectation is supported by the uncommonness of vultures in Ngorongoro in general, and at carcasses there in particular. Vultures were involved in the consumption of only one of 21 medium-sized adult carcasses observed in Ngorongoro, but on 65% of those in the Serengeti. Vultures are easily dominated and excluded by two or more hyenas at a carcass, such that the exclusive scavenging adaptation of vultures is viable only in contexts where competition with hyenas is sufficiently low to permit their frequent exclusive access to carcasses during at least a portion of the carcasses' resource life. The small number of vultures in Ngorongoro attests to the rarity of these conditions there, while low hyena to carcass ratios in the Serengeti permit vulture numbers to be extremely high (see Houston 1972, 1979 for excellent discussions on the exclusive scavenging adaptations of Serengeti vultures).

Carcass Persistence in the Serengeti and Ngorongoro: The different predator to prey ratios in Ngorongoro and the Serengeti are associated with distinct differences in the duration over which carcasses persist as food resources. Table 4.10 shows the mean persistence in hours of the three size classes of carcasses common to the Serengeti and Ngorongoro samples. Among medium-sized adults, the carcass type observed most frequently in both ecosystems, carcasses persisted almost four times longer in the Serengeti than in Ngorongoro. The one buffalo carcass discovered in the Crater persisted for less than half the average time of the larger sample from the Serengeti. Medium-sized juveniles are shown to persist for approximatley the same amount of time in both ecosystems. The reasons for this are unclear, but could well be an artifact of the method used to estimate time of death, which in both ecosystems was assumed to be midnight. Given the pattern seen among medium-sized adults, it is probable that most medium-sized juvenile carcasses discovered unfinished in the Crater were kills of the hour just before daybreak. Similar biases may be operating for the large carcasses, but their overall longer persistence than smaller carcasses (Table 4.1) reduces the impact of this source of error. The longer persistence of carcasses in the Serengeti clearly results from its lower hyena to carcass ratios, and the consequent reduced rate at which incompletely consumed carcasses are encountered by hyenas. In Ngorongoro, on the other hand, I never observed a carcass to be abandoned by non-hyaenid initial consumers without hyenas being present to scavenge immediately most if not all of the remains. With few if any partially eaten carcasses in Ngorongoro experiencing an interval of unattendence by hyenas, carcass persistence will always be relatively short.

Carcass Completeness in the Serengeti and Ngorongoro: A very strong impression from observations in Ngorongoro was the extreme rarity of hyenas finally abandoning even the most marginal head contents. In the Serengeti, on the other hand, these and other carcass parts were commonly observed

Table 4.10: Mean persistence in hours of carcasses in the Serengeti and Ngorongoro. Three carcass size classes common to samples of both ecosystems are compared. The similar persistence of medium-sized juveniles in the two ecosystems is probably an artifact of the method by which time of death was estimated (see text).

	PERSISTENCE (hrs.)				
	SERENGETI		NGORONGOGO		
Size Group	n	mean + S.D.	n	mean + S.D.	p values*
Medium-sized juveniles	8	6.6 ± 2.8	2	6.0 ± 3.0	0.44
Medium-sized adults	55	24.8 ± 28.4	16	7.0 ± 7.0	0.0001
Large adults	6	108.2 ± 62.7	1	52.0	0.29

* Mann-Whitney U-test results, with direction predicted.

to survive one or two hyena feeding bouts. The impression is supported by the high proportion of carcasses (65%, or 17 of 26) I observed to be completely consumed in Ngorongoro compared to the much lower proportion (23%, or 62 of 206 carcasses) for the Serengeti. It is also consistent with the drastically lower completeness of carcasses on final inspection in Ngorongoro (Table 4.11). The mean total completeness values for three carcass size classes on final inspection in Ngorongoro are all near or below 5%, indicating the remains to be restricted usually to frontal and mandibular pulp. Completeness values on final inspection in the Serengeti are comprised, on average, not only of all head contents, but substantial amounts of marrow as well.

Both indications of relatively thorough carcass consumption in Ngorongoro, however, are biased for methodological reasons directly related to the actual much more rapid consumption of carcasses in Ngorongoro. During any morning's observations in the Crater, I was much more likely to observe complete carcass comsumption. Situations arose in the Serengeti, on the other hand, that prevented me from monitoring the actual final abandonment of carcasses simply because of their long persistence and the associated long intervals during which incompletely consumed carcasses were unattended by scavengers. Hence, the differences in the overall completeness of carcass consumption presented above are artificially divergent, and to an uncertain degree more similar. The data serve only to corroborate the evidence presented above for extremely short carcass persistence in Ngorongoro.

There is, however, a more accurate indication that carcasses are ultimately abandoned in a less complete state in Ngorongoro than the Serengeti. Inventories of the body parts represented by weathered bones distributed on the landscape were taken on several exhaustive transects in Ngorongoro and the Seronera study area of the Serengeti. These data are similar in kind to those collected by Behrensmeyer (Behrensmeyer 1983, Behrensmeyer and Boaz 1980) in Amboseli National Park. They show the diversity of skeletal elements surviving hyena gnawing to be higher in the Serengeti than in Ngorongoro. For instance, complete or epiphyseal limb elements account for 25% (49 of 194) of all weathered elements encountered on the Seronera transects, but only 12% (22 of 186) of those in Ngorongoro. At least one epiphysis of a limb bone is destroyed by hyenas when extracting marrow. The low occurrence of limb epiphyses in Ngorongoro indicates that carcass consumption proceeded past the marrow bone stage of the consumption sequence more frequently than in the Serengeti. Among crania, typically the last parts destroyed by hyenas, only 24% (7 of 29) remained with the brain case intact in Ngorongoro while 52% (12 of 23) did in Seronera. These data confirm what my observations on fresh carcasses could not, namely, that carcasses are finally abandoned by vertebrate carnivores in

Table 4.11: Total carcass percentage indexed completeness on final inspection in the Serengeti and Ngorongoro. Completeness values do not necessarily correspond to the quantity of tissues finally abandoned by vertebrate carnivores. The values do show the greater likelihood of carcasses of each size group being observed to be completely consumed in Ngorongoro than in the Serengeti.

	TOTAL CARCASS COMPLETENESS			
Carcass	SERENGETI		NGORONGORO	
Size Group	n	mean ± S.D.	n	mean ± S.D.
Medium-sized juveniles	22	36.3 ± 39.3	4	5.5 ± 5.5
Medium-sized adults	135	29.6 ± 37.2	21	3.5 ± 8.6
Large adults	6	18.6 ± 6.5	1	4.0

a more complete condition in the Serengeti than in Ngorongoro.

Carcass completeness on abandonment by the same consumer or like series of consumers should differ between the two ecosystems only if the number of consumers, hence group appetites, differ. Kruuk (1972) notes significantly larger average hyena feeding group sizes in Ngorongoro than in the Serengeti. For wildebeest and zebra carcasses, Kruuk found average feeding group sizes to be less than 15 individuals in the Serengeti, and more than 20 in Ngorongoro. On the carcasses I observed, however, the range of hyena feeding group sizes between the two ecosystems was similar, and, given my sample, their average size was statistically indiscernible (Table 4.12). The average feeding group size of hyenas as initial consumers of medium-sized adults is, in fact, larger in the Serengeti, while the average size of feeding groups of hyenas as scavengers is only slightly larger in Ngorongoro. The average hyena feeding group size on all medium-sized adults for which data are available is just over ten individuals in both the Serengeti and Ngorongoro.

The slight differences in hyena feeding group sizes observed correspond directly to the completeness by which carcasses were observed to be consumed by hyenas (Table 4.13). The completeness of medium-sized adult lion kills scavenged by hyenas is slightly higher in Ngorongoro than in the Serengeti (Table 4.13), but this corresponds to a lower average number of hyenas per carcass in Ngorongoro (Table 4.12). The completeness on abandonment of medium-sized hyena kills in the Crater is slightly lower than in the Serengeti, but this is associated with slightly higher hyena feeding group sizes. A final illustration of the dependency of completeness of consumption on feeding group sizes is provided by episodes of hyena feeding that permitted a subsequent feeding opportunity for vultures. This circumstance occurred only once in Ngorongoro but 35 times in the Serengeti, and always resulted in carcass consumption atypically incomplete for hyenas. These high values of completeness (Table 4.13) correspond to significantly lower hyena feeding group sizes (6.4 per carcass) than are typical for the Serengeti (over 10 per carcass, Table 4.12).

The above carcass group very incompletely consumed by hyenas and scavenged by vultures reveals a context in the Serengeti in which competition by hyenas is low and the resulting feeding opportunity for other scavengers high. The rarity of this context in Ngorongoro is responsible for the observation that only 24% (5 of 21 for which data are available) of medium-sized adult carcasses persisted as food resources after hyena feeding in Ngorongoro, while 60% (72 of 121 carcasses) survived hyena feeding in the Serengeti. Contexts of low scavenging pressure by hyenas in the Serengeti has a basis in seasonal variations in levels

Table 4.12: Average feeding group sizes of hyenas as initial consumers or scavengers on medium-sized adult carcasses in the Serengeti and Ngorongoro. No consistent patterning between ecosystems is seen. See text for a discussion of the low average hyena feeding group size in the "hyenas and vultures" category for the Serengeti. Compare these data to the completeness of consumption by hyenas presented in Table 4.13. No inter-ecosystem differences in feeding group sizes within individual consumer categories are significant to the .05 level (Student t-test, 2-tailed probabilities).

	FEEDING GROUP SIZE			
	SERENGETI		NGORONGORO	
Consumer(s)	n	mean ± S.D.	n	mean ± S.D.
Hyenas only	24	10.7 ± 7.0	9	11.9 ± 8.6
Hyenas following lions	7	16.0 ± 8.6	10	9.5 ± 10.1
Hyenas and vultures	11	6.4 ± 8.6	0	---
Total	42	10.4 ± 8.3	19	10.6 ± 9.5

Table 4.13: Completeness of medium-sized adult carcasses on abandonment by various consumers in the Serengeti (A) and Ngorongoro (B). Compare like consumer categories between the two ecosystems. No differences between ecosystems in total carcass completeness among like pairs of consumer types are significant to the .05 level (Mann-Whitney U-test). Differences in completeness between ecosystems correspond directly to average hyena feeding group sizes (Table 4.12). The very high completeness of the "hyena and vulture" category in the Serengeti corresponds to low feeding group sizes and a special circumstance of carcass glut (see text).

		MEDIUM-SIZED ADULT PERCENTAGE INDEXED COMPLETENESS: MEAN/(S.D.)				
Consumer(s)	n	Viscera	Flesh	Marrow	Head Contents	Total
A) SERENGETI						
Hyenas only	27	1.9 (7.4)	3.9 (10.9)	8.7 (20.8)	26.4 (38.7)	8.0 (16.4)
Lions followed by hyenas once	14	0	0	2.5 (9.0)	14.7 (30.4)	2.4 (6.2)
Hyenas and vultures	35	5.4 (19.0)	19.1 (26.5)	61.4 (37.7)	87.9 (24.3)	42.3 (25.8)
B) NGORONGORO						
Hyenas only	9	0	0	7.1 (19.6)	19.4 (36.4)	4.8 (10.6)
Lions followed by hyenas	11	0	0	11.7 (24.9)	21.6 (34.3)	6.4 (12.5)
Hyenas and vultures	1	0	0	0	87.5	8.8

of natural mortality and live prey abundance, as described below.

Seasonality of Scavenging Opportunity in the Serengeti

Expectations: Seasonality of rainfall and the large-scale distribution of migratory herds in the Serengeti should effect local scavenger to carcass ratios, hence scavenging opportunities, in predictable ways. Levels of rainfall correspond on a seasonal basis to levels of natural mortality among herbivores, and, on a shorter term basis, to the degree of herbivore concentration and dispersal in an area. The large-scale, ecosystem-wide distribution of migratory herds is also seasonal, resting ultimately in the periodicity of rainfall and concomitant fluctuations in forage quality.

Seasonal levels of rainfall are inversely proportional to the level of nutritional stress experienced by herbivores. High stress leads to high levels of natural mortality, such as that which occurs at the end of the dry season. Houston (1979) estimates over half of annual herbivore mortality to occur in the final three months of the dry season (August through October). With the addition of natural mortality to predator-related mortality, scavenger to carcass ratios should reach their lowest point at the end of dry periods, resulting in glut situations when scavengers experience a superabundance of food. It is at these times when vultures, as exclusive scavengers, are confronted with the highest annual food availability, and the lowest levels of competition with hyenas. Wet periods, however, induce the growth of good quality forage, which, more than water availability, maintains herbivores in good condition (Sinclair 1979a). The resulting low levels of natural mortality during wet periods leads to a high degree of competition among scavengers for the tissues abandoned on the relatively few predator kills that occur.

I was unable to monitor the expectedly highest annual levels of mortality that occur at the end of the dry season in the northern and western woodlands of the Serengeti ecosystem. Still, the dryness of the short dry season and the essential failure of the following long rainy season in 1983/84 enabled me to observe unusually high levels of stress-related natural mortality on the southern plains of the Serengeti.

The seasonal movements of migratory wildebeest, zebra and Thompson's gazelle produce dramatic seasonal changes in the availability of carcasses in any one area of the Serengeti. A general observation in the field was the consistent association of high carcass densities with high densities of migratory herds, regardless of the amount of recent rainfall. Part of the increase in carcass availability results from the influx of mobile, non-breeding predators that follow herds into an area, and the consequent addition of kills to those made by residential, breeding predators. Since mobile lions and hyenas usually occur in much smaller groups than residential prides and clans (Schaller 1972, Kruuk 1972), the additional kills should be less completely consumed and offer a better opportunity for scavengers than kills by residential predators. As well, the mere presence of large numbers of herbivores increases the chances of mortality from old age, disease or accident occurring and adding further to the scavengeable food supply.

In areas seasonally devoid of migratory animals, local herbivore numbers are very low, composed of residential herbivores only. Residential breeding predators experience their most stressful season at these times (Hanby and Bygott 1979, Kruuk 1972, Schaller 1972, Sinclair 1979b). Most deaths are likely to be predator kills and these are likely to be thoroughly consumed. Any remains should be subject to very rapid and thorough scavenging by hyenas at these times.

Vultures are well adapted to conditions of seasonal fluctuations in carcass abundance. Their efficient flight allows them to travel long distances daily to track the movement of migratory herds over the whole annual round. Vulture to carcass ratios might therefore be expected to be relatively constant throughout the year compared to the expected variability in hyena to carcass ratios. The effect of relative herbivore abundance on scavenging opportunities might therefore be expected to be felt less on flesh, the food of vultures, than on marrow and head contents that are consumed mostly by hyenas.

Smaller-scale differences in the local density distribution of herbivores resulting from short term variations in rainfall amounts should also effect scavenging opportunities. Regular rainfall replenishes small temporary pools of water that enable herbivores to be dispersed. Dry conditions, however, quickly reduce water availability to few sources around which herbivores are forced to concentrate. Such short term and small scale concentration and dispersal was often observed in the Seronera study area during periods of irregular rainfall. Since the ranges of residential predators are fixed by territorial boundaries (e.g. Bertram 1979, Hanby and Bygott 1979, Kruuk 1972, Schaller 1972), the degree of herbivore concentration should be inversely proportional to the number of consumers at any one carcass. A higher local concentration of carcasses should therefore result in less complete carcass consumption, and better scavenging opportunities.

The co-effects of relative rainfall and herbivore abundance in the Serengeti should produce larger variations in scavenging opportunity than that resulting from any one effect alone. Scavenger to carcass ratios and competition for carcasses should be lowest during dry periods of high natural mortality and herbivore concentration, and in areas temporarily supporting large numbers of migratory ungulates. This situation characterizes the northern and western woodlands during the long dry season, and the period which I observed in February through April in the southern plains. Conversely, the highest scavenger to carcass ratios and competition should occur during wet periods in areas supporting only small numbers of residential herbivores, such as during the rainy season in the woodlands. Houston (1979), assuming 2% annual mortality, estimated the density of mortality for the seasonal habitats of the Serengeti to be as follows: 1 carcass/33 km^2 on the plains during the wet season, 1/300 km^2 on the plains during the dry season, and 1/412 km^2 in the woodlands during the wet season. Houston does not provide an estimate of carcass densities for the woodlands in the dry season, but it must approach if not exceed that for the wet season on the plains owing to high levels of natural mortality.

<u>Stratifying the Carcass Sample Seasonally</u>: Daily records were kept on the incidence and intensity of showers in each study area being monitored. The most complete records are those from the intensively monitored Seronera study area. In study areas visited infrequently and for short durations, shower data were collected from reports of other researchers recently in the area of interest, and by the conditions I encountered upon visiting the area, including the presence or absence of temporary pools of water, the color of the grass, and the amount of dust. Each day of carcass search could thereby be classified into relatively wet or relatively dry days.

Dry days are those following a week in which at most one or two light showers fell. This amount of rainfall always proved insufficient to produce temporary surface accumulations of water. Most days classified as dry were preceded by at least one week of no rain producing extremely dry, dusty conditions, which, if persisting, turned the grass brown. Wet days are those on which at least one heavy shower or several continuous days of light showers fell during the preceding week. This level of rainfall usually produced many temporary pools of water and a flush of green on previously brown expanses of grass.

The classification of wet and dry days does not necessarily correspond to the rainy and dry seasons. This is a necessary consequence of the lack of pronounced seasonality in rainfall during the year encompassing the field study. Some dry days occurred within a period of relatively heavy rainfall, and occasionally wet days occurred in predominantly dry periods. As a result, the rainfall classification used is not highly correlated with expected levels of natural mortality. Of the 27 certain natural deaths encountered in the Serengeti, 19 occurred on dry days, but eight of these were drowned wildebeest. I encountered 8 natural deaths on wet days, which is not substantially lower than the 11 unaccidental deaths observed on dry days.

The rainfall classification used, however, is sensitive to changes in local density distributions of herbivores. Such changes are rapid responses to the availability of water sources. Whether the mechanism is the degree of local herbivore dispersion or the level of natural mortality, rainfall-based variations in scavenging opportunity should have a similar effect for both.

The presence or absence of migratory herds in a study area was also monitored on a daily basis. Influxes of migratory herds to a study area were extremely conspicuous, reaching abundances of tens of thousands of animals in the relatively small area. The departure of the migratory herds left an area devoid of all but small numbers of Grant's gazelle, topi, hartebeest, buffalo, giraffe and warthog. Days of herbivore abundance were those when very large concentrations of wildebeest and zebra, whose migratory movements are in concert, were present. Thompson's gazelle are more dry-adapted than wildebeest and zebra and have somewhat different seasonal movements (e.g. Maddock 1979). An area supporting many tommies but few or no larger migrants was considered to be relatively devoid of animals. Most days of herbivore scarcity, however, were those during which only small numbers of residential ungulates were present.

Seasonality of Carcass Availability: The effect of seasonal variations on the availability of carcasses is indicated by data in Table 4.14. The data describe the average number of carcasses discovered per day of relative rainfall and/or herbivore abundance. The number of carcasses discovered per day disregards differences in daily searching distance, which ranged from 30 to 80 linear kilometers. The data also are not a measure of the absolute number of carcasses present per day in a given study area, as surely some carcasses, particularly small ones, were never discovered. Still, daily carcass yields almost always corresponded to my sense of the potential availability of carcasses, and the data in Table 4.14 are largely consistent with the expectations described above.

Most carcasses in the Serengeti were discovered on days and in areas supporting large concentrations of wildebeest and zebra. An average of 2.9 carcasses were discovered on such days of herbivore abundance compared to only 1.1 per day on days of prey scarcity. The difference is highly significant. Among days of herbivore abundance, a significantly higher number of carcasses were discovered on dry days than on wet days. However, relative amounts of rainfall alone, according to my classification, do not seem

Table 4.14: Average number of carcasses discovered per day in the Serengeti during periods of different rainfall (dry vs. wet) and/or live herbivore abundance (many vs. few), and in Ngorongoro. Using Student's T-test, all paired differences within the Serengeti are significant to at least the .05 level, except for dry/few vs. wet/few, and all wet vs. all dry. Ngorongoro is significantly different from the whole of the Serengeti, and each of the combined rainfall/herbivore abundance categories except for wet/few.

Seasonal Context	Days of Searching	Carcasses Discovered	Carcasses/Day mean + S.D.
Dry/Many	24	84	3.5 ± 1.9
Dry/Few	22	20	0.9 ± 1.5
Wet/Many	36	92	2.6 ± 1.9
Wet/Few	40	48	1.2 ± 1.1
All Many	60	176	2.9 ± 2.0
All Few	62	68	1.1 ± 1.3
All Dry	46	104	2.3 ± 2.1
All Wet	76	140	1.8 ± 1.7
Total Serengeti	122	244	2.0 ± 1.9
Total Ngorongoro	25	28	1.1 ± 1.0

to effect carcass availability measurably. The fundamental seasonal-based availability of carcasses in the Serengeti seems to be dependent mostly on the presence or absence of migratory herds rather than on the relative dryness of periods preceding days of carcass search. Whether this indication of the fundamental availability of carcasses corresponds to the actual, or realized, scale of scavenging opportunity, as measured by carcass completeness and persistence, will be evaluated later.

Table 4.14 also provides a comparison of the average daily carcass yield in the Serengeti to that in Ngorongoro. Compared to the Serengeti as a whole, significantly fewer carcasses were discovered per day in Ngorongoro (2.0/day vs 1.1/day, respectively). Despite the high predation rates of the Crater, the short persistence of carcasses there is apparently responsible for the low numbers remaining to be discovered on my early morning carcass searches. Indeed, it was a common (and frustrating) observation to see full-bellied lions and hyenas at dawn in the Crater, but absolutely no trace of the carcass. Note, however, the similarity in daily carcass yields from Ngorongoro and the Serengeti in contexts of herbivore scarcity (both at 1.1 carcasses discovered per day). It is precisely at times of low prey availability in the Serengeti that competition for carcasses is expected to be the highest, and apparently to a degree approaching that seen in Ngorongoro.

In summary, data on the number of carcasses discovered per day suggest that carcass persistence and completeness on abandonment should be highest in the Serengeti during periods of herbivore abundance and low rainfall. Scavenging opportunities should be the poorest during periods of prey scarcity and similar to those in the always highly competitive Ngorongoro Crater.

<u>Methodological Biases in Seasonality Data</u>: Carcass persistence data are subject to methodological biases that produce results insensitive to seasonal variations. During times of carcass abundance, only a small percentage of carcasses could be monitored until consumption was complete. Among medium-sized adult carcasses, I observed only 23% (9 of 39) until complete consumption during dry periods of herbivore abundance, but 56% (9 of 16) during wet periods of herbivore scarcity. These data do suggest that carcasses will persist longer during periods of low competition, but they are at least in part an artifact of my tendency to leave the carcasses before feeding was complete in favor of visiting another carcass in the area. Such an opportunity was more likely to arise, of course, during periods of carcass abundance. Contributing further to the bias is the fact that carcass abundance was usually encountered in study areas monitored for only one or two days at a time (Naabi-Ndutu and Simba Kopjes). This reduced my chances of revisiting a carcass before leaving the study area. Furthermore, the Naabi-Ndutu and Simba Kopjes study

areas are primarily open plains habitats devoid of features on the landscape permitting easy relocation of carcasses. Hence, of the 30 medium-sized adult carcasses discovered during periods of prey abundance and remaining with edible parts on final inspection, 63% occurred in the two plains study areas. At the other extreme, during periods of carcass scarcity (wet, few live herbivores present) only one of the seven carcasses in this sample is from outside of intensively monitored Seronera. The effect of these biases is to reduce artificially the observed persistence of the average carcass during periods of carcass abundance. The few carcasses observed in this context until the end of their resource life were, for one reason or another, very quickly consumed relative to the majority of carcasses for which persistence could not be calculated. The available, biased data show carcass persistence during periods of carcass abundance to be slightly <u>lower</u> than during periods of prey scarcity. I could conceive of no way to correct for the contributing biases.

Data on seasonality of carcass completeness are not subject to the above biases. Comparisons of carcass completeness are made within groups of like consumers that were <u>observed</u> to abandon carcasses with edible tissues. Such abandonments occur over a much shorter time scale than does complete carcass consumption which is the basis for measures of persistence. If, owing to logistical considerations, the initial or later abandonment of a carcass was not observed, that carcass is necessarily excluded from analyses of carcass completeness.

<u>Seasonality of Carcass Completeness</u>: The completeness by which medium-sized adult carcasses are abandoned by like consumers in different contexts of rainfall and herbivore abundance is shown in Tables 4.14 - 4.19. The tables include data on the average number of consumers primarily responsible for carcass consumption if these could be accurately determined. Note again that sample sizes for which feeding group size was determined are sometimes smaller than those for the corresponding completeness data because I was not always able to observe consumption but could nonetheless unambiguously identify the consumer(s) responsible. The number of carcass consumers was shown previously in comparisons between the Serengeti and Ngorongoro to be an important intervening variable conditioning carcass completeness, as it is shown to be here in different seasonal contexts.

The total completeness of medium-sized adult carcasses consumed by lions (Table 4.15) and lions followed by vultures (Table 4.16) shows only minor differences among contexts of different rainfall and herbivore abundance. Greater distinctions, however, are seen in the completeness of viscera and particularly flesh, the only tissues consumed by lions and vultures on wildebeest and zebra adults. For both consumer categories, relative herbivore

Table 4.15: Completeness on abandonment of medium-sized adult lion kills in different seasonal contexts in the Serengeti. Seasonal contexts include relative rainfall (wet vs. dry) and herbivore abundance (many vs. few). Also shown are the average lion feeding group sizes. Note the negative correlation between feeding group sizes and the completeness by which flesh, the most variable component, is abandoned among most combinations of relative rainfall and herbivore abundance. The completeness by which lions consume carcasses seems to be more dependent on relative herbivore (prey) abundance.

	MEDIUM-SIZED ADULT PERCENTAGE INDEXED COMPLETENESS: MEAN/(S.D.)						# lions/carcass	
Seasonal Context	n	Viscera	Flesh	Marrow	Head Contents	Total	n	mean/(S.D.)
Dry/Many	5	2.5 (5.0)	13.9 (19.3)	100.0	98.8 (2.5)	56.2 (9.0)	5	8.4 ± 6.2
Dry/Few	1	0	6.3	100.0	93.8	51.3	1	15.0
Wet/Many	8	4.7 (12.4)	18.0 (18.5)	99.6 (1.0)	94.5 (7.3)	58.7 (8.8)	8	5.5 ± 2.3
Wet/Few	5	0	2.0 (2.2)	100.0	92.5 (6.1)	50.3 (1.4)	4	5.0 ± 2.4
All Many	13	3.8 (10.3)	16.4 (18.9)	99.8 (0.8)	96.1 (6.3)	57.8 (9.0)	13	6.6 ± 4.5
All Few	6	0	2.7 (2.6)	100.0	92.7 (5.6)	50.5 (1.3)	5	7.0 ± 4.6
All Dry	6	2.1 (4.7)	12.6 (17.9)	100.0	97.9 (2.9)	55.4 (8.4)	6	9.5 ± 6.2
All Wet	13	2.9 (10.0)	11.8 (16.5)	99.8 (0.8)	93.8 (6.9)	55.5 (8.1)	12	5.3 ± 2.4

abundance seems to have a stronger influence on the amount of flesh abandoned than do relative amounts of rainfall.

Lions tend to abandon carcasses with substantial quantities of flesh on the head, distal limbs and neck during periods of prey abundance (Table 4.15). Prey scarcity, however, is associated with the consumption by lions of virtually all flesh, with the occasional exception of maxillary flesh and more rarely flesh on the extreme distal portion of limbs. Viscera are also always completely consumed at these times, while small amounts remain during periods of prey abundance. Within contexts of prey abundance, slightly more viscera and flesh are abandoned during wet periods.

The completeness of flesh consumption from medium-sized adults by lions seems to be dependent in part on the size of their feeding groups (Table 4.15). Feeding group size is inversely proportional to the completeness of flesh abandoned by lions for all four seasonal categories save that of prey scarcity during wet periods. The latter context is that during which flesh consumption tends to be most complete, but for which the smallest average group size of lions was observed. The reason for this unconformity is not known, but it is responsible for the lack of correlation between average feeding group size and flesh completeness when relative prey abundance or relative rainfall are considered alone (Table 4.15). The lack of a complete correlation is curious in light of expectations that the presence of migratory herds is usually associated with an increased likelihood of encountering lion kills made by small groups of mobile, non-breeding males that follow the herds (cf. Schaller 1972, Hanby and Bygott 1979). Lion kills of medium-sized adults during periods of migratory herd absence, on the other hand, are more likely to be made exclusively by large residential breeding prides which should more completely consume carcasses regardless of the amount of rainfall and consequent dispersion of prey.

Relative prey abundance rather than relative amounts of rainfall also seems to have a greater effect on the completeness by which vultures scavenge lion kills (Table 4.16). The pattern, however, is reversed from that seen for lions alone. Flesh abandoned by lions is more completely scavenged by vultures during periods of prey abundance than when few prey are in the area. In the latter context, scavenged lion kills are abandoned by vultures with flesh remaining partially eaten over most of the forequarters. This pattern can be explained by the strong correlation between prey abundance and vulture abundance, a phenomenon caused by the ability of most vultures, nesting or not, to follow migratory herds over large areas. Vultures are extremely rare in areas of the Serengeti supporting only residential ungulates, and appear at carcasses in very small numbers. An influx to the area of migratory herds,

Table 4.16: Completeness of medium-sized adult lion kills after scavenging by vultures in different seasonal contexts of relative rainfall (wet vs. dry) and live herbivore abundance (many vs. few). Note the virtual complete scavenging of viscera and flesh by vultures in contexts of herbivore abundance from the substantial amounts left by lions (Table 4.15).

		MEDIUM-SIZED ADULT PERCENTAGE INDEXED COMPLETENESS: MEAN/(S.D.)				
Seasonal Context	n	Viscera	Flesh	Marrow	Head Contents	Total
Dry/Many	5	0	2.5 (3.3)	100.0	95.0 (7.3)	50.6 (2.0)
Dry/Few	2	0	27.1 (27.1)	100.0	96.9 (3.1)	61.9 (12.5)
Wet/Many	9	0	5.7 (5.7)	99.3 (1.9)	94.5 (5.5)	52.3 (3.3)
Wet/Few	1	0	0	100.0	81.3	48.1
All Many	14	0	4.5 (5.2)	99.6 (1.6)	94.7 (6.2)	51.7 (3.0)
All Few	3	0	18.1 (25.6)	100.0	91.7 (7.8)	57.3 (12.1)
All Dry	7	0	9.5 (18.5)	100.0	95.6 (6.4)	53.9 (8.6)
All Wet	10	0	5.1 (5.7)	99.4 (1.9)	93.1 (6.5)	51.9 (3.3)

however, is accompanied by an extremely conspicuous and dramatic increase in vulture numbers. Lion kills draw large groups of vultures at these times, which swamp and voraciously consume most remaining flesh and viscera immediately upon abandonment by lions.

In summary, the relatively large amounts of flesh observed to be abandoned by lions during periods of prey abundance will be quickly and thoroughly scavenged by vultures. Vultures have a lesser impact as scavengers in areas supporting mostly residential herbivores, but in this context, relatively large lion feeding group sizes result in the infrequent abandonment of any flesh. Unattended lion kills of medium-sized adults will therefore provide little to no flesh at all times of the year in the Serengeti, with further feeding opportunities being restricted to tissues within bones.

Hyena kills of medium-sized adults are consumed in a pattern similar to that observed for lion kills (Table 4.17). Although hyenas always consume their kills much more thoroughly than lions, total and component part completeness show the greatest differences within the context of relative prey abundance. Relative rainfall has a smaller but still measureable effect. During periods of prey scarcity, carcasses are abandoned by hyenas with only marginal head contents; hyenas are more likely to leave not only head contents and marrow, but some flesh and viscera in contexts of prey abundance. Wet periods, too, witness hyenas abandoning carcasses relatively complete, while kills made on dry days are abandoned with little marrow and head contents, in a pattern similar to that seen in Ngorongoro. This pattern is unexpected in light of the higher expected levels of natural mortality during dry periods. The carcasses comprising the whole sample in Table 4.17, however, were all deemed to be hyena kills that occurred in Seronera. Natural mortality was not observed in Seronera to the extent that it was on the plains to the south. Prey abundance in Seronera was usually associated with relatively wet periods that drew large herds of wildebeest and zebra from the north to the margin of the plains. In this context, hyena feeding group sizes were on average small, resulting in relatively incomplete consumption.

Figure 4.4 shows the relationship between mean hyena feeding group sizes and total carcass completeness on abandonment. The two variables are inversely proportional. A similar pattern is seen for head content and marrow completeness as well. The figure clearly separates total carcass completeness along lines of relative prey abundance, with thorough consumption occurring during periods of prey scarcity when average feeding group sizes of 13 or more were observed. Large group size in contexts of prey scarcity is consistent with the expectation that mostly large, residential clans will occur in areas

Table 4.17: Completeness on abandonment of medium-sized adult hyena kills in different seasonal contexts in the Serengeti, and in Ngorongoro. Average hyena feeding group sizes are also shown. Note the strong negative correlation between feeding group sizes and the completeness of carcass consumption (see Figure 4.4). Greatest differences in carcass completeness result from differences in herbivore abundance, though relative rainfall is also important.

		MEDIUM-SIZED ADULT PERCENTAGE INDEXED COMPLETENESS: MEAN/(S.D.)					# Hyenas/Carcass	
Seasonal Context	n	Viscera	Flesh	Marrow	Head Conts.	Total	n	mean $\pm$ S.D.
Dry/Many	8	0	0	5.3 (9.0)	24.2 (35.7)	4.5 (7.1)	8	9.9 $\pm$ 8.1
Dry/Few	2	0	0	0	0	0	2	21.5 $\pm$ 3.5
Wet/Many	13	3.8 (10.3)	8.1 (14.6)	14.9 (27.7)	34.1 (44.7)	13.2 (21.7)	10	8.2 $\pm$ 3.5
Wet/Few	4	0	0	0	18.8 (20.7)	1.9 (2.1)	4	13.0 $\pm$ 7.1
All Many	21	2.4 (8.3)	5.0 (12.1)	11.2 (23.0)	30.4 (41.8)	9.9 (18.2)	18	8.9 $\pm$ 6.0
All Few	6	0	0	0	12.5 (19.1)	1.3 (1.9)	6	15.8 $\pm$ 7.4
All Dry	10	0	0	4.2 (8.3)	19.4 (33.4)	3.6 (6.6)	10	12.2 $\pm$ 8.7
All Wet	17	2.9 (9.1)	6.2 (13.2)	11.4 (25.0)	30.5 (40.9)	10.5 (19.6)	14	9.6 $\pm$ 5.3
Ngorongoro	9	0	0	7.1 (19.6)	19.4 (36.4)	4.8 (10.6)	9	11.9 $\pm$ 8.6

Figure 4.4: Relationship between mean hyena feeding group size and the completeness by which their kills of medium-sized adults are abandoned in different seasonal contexts in the Serengeti. Comparable data are shown for Ngorongoro, which conform well to the good negative correlation seen for the Serengeti sample. Higher feeding group sizes and more complete carcass consumption are associated with prey scarcity in the Serengeti. Accompanying data appear in Table 4.17.

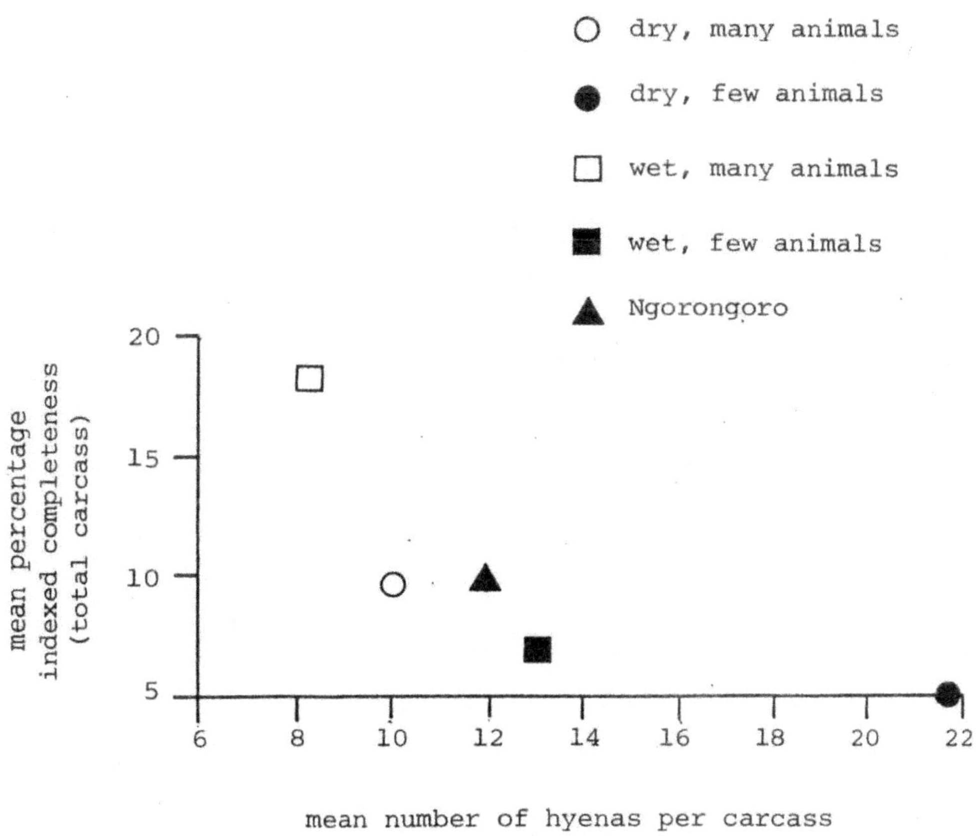

occupied only by residential ungulates. With the migration of large herds of wildebeest and zebra into Seronera come solitary or paired mobile hyenas which are likely to feed on carcasses in smaller groups. Average group sizes during periods of prey abundance are under ten, resulting in the relatively incomplete consumption of the average carcass. Hyena kills in Ngorongoro are shown to be consumed by groups of intermediate average size, and to be abandoned with a completeness that conforms well to the seasonal pattern described by the Serengeti sample.

The season-specific completeness of medium-sized adult lion kills scavenged by hyenas is shown in Table 4.18. The usually minimal scavenging by vultures (Table 4.16) is included to an unindicated degree.

Greater differences in the intensity of hyena scavenging are associated with relative amounts of rainfall rather than relative prey abundance. This pattern is the converse of that seen for lions on their kills, vultures scavenging lion kills, <u>and</u> hyenas on their kills. Hence, the thoroughness of hyena scavenging is greatest during wet periods, regardless of the abundance of prey. The complete series of marrow bones and head contents abandoned by lions are virtually all scavenged by hyenas, while during dry periods, hyena scavenging is less thorough.

As with other consumer categories, the thoroughness of hyena scavenging is inversely proportional to the number of hyenas in the scavenging group. Figure 4.5 shows the relationship for total carcass completeness only, but it also holds for the completeness by which marrow and head contents are scavenged. Ngorongoro clusters closely with the wet sample from the Serengeti, where scavenging group sizes are close to or exceed ten individuals, and total carcass completeness is reduced to roughly 6%. The pattern in the Serengeti might be explained by the greater dispersion of prey during wet periods, which produces fewer lion kills per hyena territory, and a larger number of hyenas to scavenge each abandoned carcass. The pattern does not seem to result from higher rates of natural mortality and the consequent lower competition for, and completeness on abandonment of, carcasses during dry periods. If this were the case, one would expect the head contents of natural deaths initially consumed by lions to be preferred over marrow bones by scavenging hyenas in a pattern not unlike that seen in subadults at all times. This results from the depletion of fat in marrow bones of nutritionally stressed ungulates which, if severe enough, leads quickly to death (e.g. Harris 1945).

Hyenas and vultures comprise the final set of consumers to be considered in the analysis of seasonal variations in carcass completeness (Table 4.19). This sample is drastically different from others involving hyena consumption for a number of reasons. 31 of the 34 carcasses

Table 4.18: Completeness of medium-sized adult lion kills after scavenging by hyenas and occasionally vultures in different seasonal contexts in the Serengeti, and in Ngorongoro. Average scavenging group sizes of hyenas are shown, and are negatively correlated to carcass completeness on abandonment (Figure 4.5). Intensity of hyena scavenging seems to be more dependent on relative rainfall than on herbivore abundance.

		MEDIUM-SIZED ADULT PERCENTAGE INDEXED COMPLETENESS: MEAN/(S.D.)					# Hyenas/Carcass	
Seasonal Context	n	Viscera	Flesh	Marrow	Head Conts.	Total	n	mean + S.D.
Dry/Many	11	0	3.0 (6.4)	13.5 (23.9)	64.8 (36.9)	13.0 (13.5)	3	7.3 ± 4.9
Dry/Few	9	0	6.9 (19.6)	24.5 (30.0)	61.8 (44.1)	18.4 (21.4)	1	5.0
Wet/Many	25	0	0.3 (0.8)	6.1 (16.5)	28.3 (38.9)	5.2 (8.4)	8	10.3 ± 8.0
Wet/Few	9	0	0	6.7 (19.0)	24.3 (32.1)	4.7 (8.3)	6	12.5 ±10.5
All Many	36	0	1.1 (3.8)	8.4 (19.3)	39.4 (41.8)	7.6 (10.9)	11	9.5 ± 7.4
All Few	18	0	3.5 (14.3)	15.6 (26.7)	43.1 (42.9)	11.5 (17.6)	7	11.4 ±10.1
All Dry	20	0	4.8 (14.2)	18.4 (27.4)	63.5 (40.3)	15.4 (17.7)	4	6.8 ± 4.4
All Wet	34	0	0.2 (0.7)	6.3 (17.2)	27.2 (37.2)	5.3 (8.4)	14	11.2 ± 9.2
Ngorongoro	11	0	0	11.7 (24.9)	21.6 (34.3)	6.4 (12.5)	10	9.5 ±10.1

Figure 4.5: Relationship between hyena scavenging group size and the completeness by which they abandon medium-sized adult lion kills in different seasonal contexts in the Serengeti. Comparable data for Ngorongoro are shown to conform to the good negative correlation among the Serengeti samples. Larger scavenging groups and more thorough scavenging are associated with relatively high amounts of rainfall, rather than prey scarcity alone. Accompanying data appear in Table 4.18.

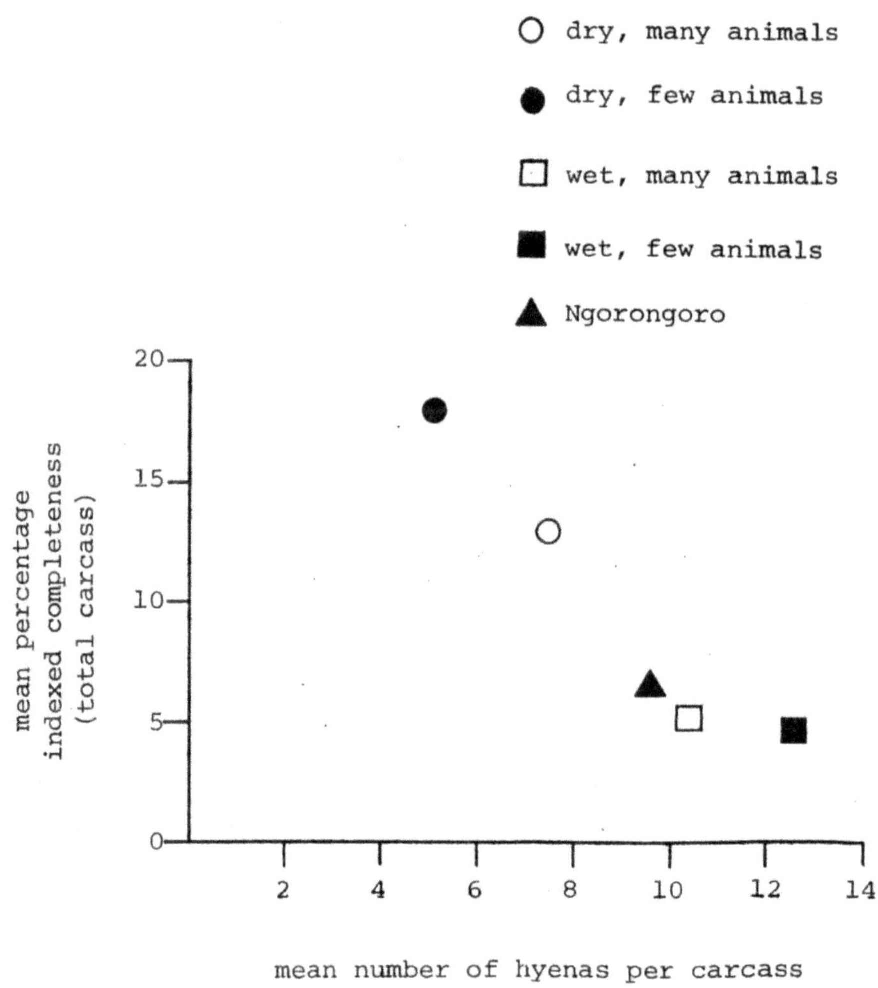

Table 4.19: Completeness by which hyenas and vultures scavenge natural deaths of medium-sized adult carcasses in different seasonal contexts in the Serengeti. Only 4 of the carcasses for which cause of death could be determined were probable hyena kills. More than half of the carcasses occurred in dry periods, and all but three in areas of herbivore abundance. The high completeness of carcasses on abandonment by hyenas is associated with small scavenging group sizes, as is expected to occur when scavengers are saturated with carcasses.

Seasonal Context	n	MEDIUM-SIZED ADULT PERCENTAGE INDEXED COMPLETENESS:MEAN/(S.D.) Viscera	Flesh	Marrow	Head Conts.	Total	# Hyenas/Carcass n	mean ± S.D.
Dry/Many	19	4.6 (19.5)	25.9 (27.1)	65.5 (38.9)	92.8 (19.1)	47.5 (27.0)	5	4.0 ± 3.6
Dry/Few	2	0	11.1 (9.7)	89.1 (11.0)	65.7 (34.4)	47.2 (3.4)	0	---
Wet/Many	12	8.3 (20.6)	13.0 (26.4)	56.1 (35.2)	83.3 (28.2)	37.0 (24.7)	5	4.0 ± 1.9
Wet/Few	1	0	0	0	93.8	9.4	0	---
All Many	31	6.0 (20.0)	20.9 (27.5)	61.9 (37.8)	89.1 (23.5)	43.5 (26.6)	10	4.0 ± 2.9
All Few	3	0	7.4 (9.5)	59.4 (42.9)	75.0 (31.0)	34.6 (18.0)	0	---
All Dry	21	4.2 (18.6)	24.5 (26.3)	67.7 (37.8)	90.2 (22.5)	47.5 (25.7)	5	4.0 ± 3.6
All Wet	13	7.7 (19.9)	12.0 (25.6)	51.8 (37.0)	84.1 (27.3)	34.9 (24.8)	5	4.0 ± 1.9

in this sample occurred during periods of prey abundance, and 25 were discovered on the Simba Kopje and Naabi-Ndutu plains during the very dry period of January through March when the migratory herds were extremely concentrated. Of the 22 carcasses for which the cause of death could be confidently determined, 18 were natural deaths and only four were probable hyena kills. Most carcasses were discovered first by vultures and only subsequently by hyenas. These circumstances describe the times of greatest carcass abundance encountered during my study period, when as many as eight very complete natural deaths were discovered on any single day of searching. Carcass abundance was sufficiently high at these times to produce a glut for scavengers, possibly on the order of that expected for the northern and western woodlands during the end of the dry season.

The main consequence of the carcass glut is an atypically high completeness by which hyenas abandon carcasses to other scavengers. Consumption by hyenas often did not proceed beyond viscera or small amounts of hindquarter flesh. Most of the remaining flesh was consumed by vultures, whose effect on reducing carcass completeness to the values seen in Table 4.19 is probably equal to or greater than that of hyenas. The small amount consumed by hyenas is directly related to a small average feeding group size of approximately four individuals. In this context, I frequently observed hyenas foraging and feeding either singly or in pairs. The small feeding group sizes are not the result of low hyena densities, but rather of the large number of available carcasses. Many of the carcasses occurred on the Naabi-Ndutu plains in the vicinity of at least two active hyena dens. I observed on several occasions one or two hyenas feeding from a recent natural death while the rest of the clan lay full-bellied and uninterested a short distance away. These data illustrate nicely how low hyena to carcass ratios can lead to good feeding opportunities for non-hyaenid scavengers. Carcass glut, though occurring in the short dry season during my study, should in years of more normal rainfall be restricted to the very end of the long dry season.

<u>Summary of Seasonality</u>: Seasonality in both rainfall and herbivore abundance has distinct effects on the completeness by which carcasses are abandoned to scavengers in the Serengeti. Initial consumers, whether as predators or scavengers of natural deaths, abandon carcasses of a completeness that seems to depend more on prey abundance than on relative amounts of rainfall. Despite the addition of kills made by mobile predators to those made by residential clans or prides, the associated influx of migratory herds to an area is apparently sufficient to decrease scavenger to carcass ratios. The higher availability of carcasses is associated with smaller average feeding group sizes and the initial abandonment of a greater amount of edible tissues to scavengers. Adding

the effect of dry periods stressful to herbivores to contexts of prey abundance further reduces scavenger to carcass ratios, creating the best seasonally-based scavenging opportunities. It is in fact only at these times that scavengers are likely to encounter unattended carcasses remaining with large quantities of flesh.

Hyenas as secondary consumers, seen here as scavengers of lion kills, seem instead to consume remains to a degree dependent on relative amounts of rainfall. Scavenging group size again was seen to be an intervening variable affecting the completeness of consumption. Lion kills scavenged by hyenas almost always provide the most marginal opportunity for other scavengers, which is only slightly better during wet periods than dry ones. Wet periods in general, however, were seen to offer a poorer opportunity to scavengers unless accompanied by the presence of large numbers of prey and potential carcasses.

Habitat Type and Scavenging Opportunity

A striking observation in the Serengeti was the rarity of sightings of hyenas in riparian woodland. Riparian woodlands in the Serengeti study areas include both riverine and lacustrine woodlands. Riparian woodlands are in general more densely wooded than the open *Acacia* woodland that dominates the northern and western portions of the Serengeti. Riverine woodlands frequently monitored for carcasses occur along the Seronera and Wandamu rivers of the Seronera study area (Figure 2.5). The two rivers are perennial and are bordered by narrow strips of dense to open woodland as little as 100 meters wide, and no more than three quarters of a kilometer wide. Much denser and extensive riverine woodland occurs along larger rivers in the Serengeti to the north, but these were not monitored. Denser and more extensive lacustrine woodland occurs along the flanks of and within Olduvai Gorge around perennial lakes Ndutu and Masek within the Naabi-Ndutu study area (Figure 2.6). On the handful of occasions I observed hyenas in riparian woodland, they were present in relatively small numbers (less than ten, and usually around five individuals) and conspicuously wary, apparently of encountering lions.

Hyenas, however, are very common in open plains and open *Acacia* woodland throughout the Serengeti. Hyenas will also forage without hesitation up to the margins of riparian woodland. The disinclination of hyenas to frequent denser riparian woodland is consistent with Kruuk's (1972) observation of their preference for open vegetation habitats.

These observations are supported by data on the habitat-specific frequency of hyenas and other carnivores as initial consumers (Table 4.20). Hyenas were observed or

Table 4.20: Frequency of carnivores as initial consumers in different habitats of the Serengeti, and in Ngorongoro. All carcasses are considered. Note the rarity of hyenas as initial consumers within riparian woodland, and on its margin.

	INITIAL CONSUMERS											
	Lions		Hyenas		Vultures		Cheetah		Other		Total	
Habitat	n	%	n	%	n	%	n	%	n	%	n	%
Acacia woodland	9	39.1	12	52.2	2	8.7	0	--	0	--	23	100.0
Plains	44	33.6	56	42.7	16	12.2	12	9.2	3	2.3	131	100.0
Riparian woodland margin	9	56.3	1	6.3	4	25.0	0	--	2	12.5	16	100.1
Riparian woodland proper	22	88.0	2	8.0	1	4.0	0	--	0	--	25	100.0
Total Serengeti	84	43.1	71	36.4	23	11.8	12	6.2	5	2.6	195	100.1
Ngorongoro	14	50.0	14	50.0	0	--	0	--	0	--	28	100.0

implicated as initial consumers on approximately half of all carcasses found in open plains, open *Acacia* woodland, and in Ngorongoro, which is mostly plains. On riparian woodland margins or within riparian woodlands, however, hyenas were initial consumers on less than 10% of all carcasses. Lions show a reversed pattern, accounting for the initial consumption of 88% of all carcasses found within riparian woodland, and more than half of those on its margins. Vultures are rare initial consumers in all habitats, and cheetah were initial consumers only 12 times, all on the plains. Other carnivores were rarely responsible for initial consumption. Lions are therefore the principle conditioner of scavenging opportunities in riparian woodland, while hyenas play the role in more open habitats. Given the typical complete marrow and head content remains of lion kills on their preferred, medium-sized adult prey, and the rare circumstances in which hyenas abandon more than minimal remains to scavengers, the expectation arises that much better scavenging opportunities will be encountered in riparian woodland than in more open habitats.

As major scavengers, however, hyenas have a large impact on reducing carcass yields in all habitats. Table 4.21 shows the habitat-specific frequency by which various carnivores were involved, initially or as scavengers, in carcass consumption. Hyenas ate from over 75% of all carcasses in all habitats regardless of vegetation cover, and to a degree approaching or similar to that observed in the highly competitive Ngorongoro Crater. Vultures consumed from over 50% of all carcasses in all habitats, while the figures for lion and cheetah, which were rarely if ever observed to scavenge, are similar to those for initial consumption.

The apparent contradiction between the rarity of hyena sightings and their scavenging of most carcasses in riparian woodland is resolved by considering the rapidity by which hyenas scavenge in different habitats. Table 4.22 shows such data for lion kills. The proportion of Serengeti lion kills discovered by me that had yet to be scavenged by hyenas was highest within riparian woodlands and lowest in *Acacia* woodlands. Comparable data for Ngorongoro are similar to those for *Acacia* woodland. Moreover, of the lion kills still being consumed by lions, in only one instance within riparian woodland were hyenas intent on scavenging present. On the other hand, hyenas were present and waiting to scavenge lion kills immediately in over half the cases on the plains, and in all cases in *Acacia* woodland and Ngorongoro. These data suggest there to be a longer interval between a kill by lions and its discovery by hyenas in riparian woodland than in more open habitats.

The suggestion is confirmed by data on carcass persistence presented in Table 4.23. Only medium-sized adults within the Seronera study area are considered. The

Table 4.21: Frequency by which different carnivores were involved in feeding, whether as initial consumers or scavengers, in different habitats of the Serengeti, and in Ngorongoro. All carcasses are considered. Note how hyenas consume from over three quarters of all carcasses in all habitats, despite their rarity as initial consumers in riparian woodland (Table 4.20).

		INITIAL CONSUMERS AND SCAVENGERS									
	Total	Lions		Hyenas		Vultures		Cheetah		Other	
Habitat	Carcass n	n	%	n	%	n	%	n	%	n	%
Acacia woodland	26	9	34.6	22	84.6	13	50.0	0	--	2	7.7
Plains	131	47	35.9	103	78.6	87	66.4	12	9.2	14	10.7
Riparian woodland margin	20	9	45.0	20	100.0	20	100.0	0	--	2	10.0
Riparian woodland proper	29	22	75.9	22	75.9	20	69.0	0	--	1	3.4
Total Serengeti	206	87	42.2	167	81.1	140	68.0	12	5.8	19	9.2
Ngorongoro	28	14	50.0	27	96.4	3	10.7	0	--	8	28.6

Table 4.22: Scavenging pressure by hyenas on lion kills in different habitats of the Serengeti, and in Ngorongoro. Column B refers to the proportion of all lion kills yet to be scavenged by hyenas upon my discovery of the carcass. Column C is a subset of column B, referring to the proportion of kills still being consumed by lions at which hyenas were present and waiting to scavenge on abandonment by lions. Note both the high proportion of kills yet to be scavenged and the low proportion ready to be scavenged in riparian woodland proper compared to more open habitats.

Habitat	Total # of lion kills	B Kills unscavenged by hyenas n	%	C Kills ready to be scavenged n	%
Acacia woodland	9	1	11.1	1	100.0
Plains	44	17	38.6	9	52.9
Riparian woodland margin	9	4	44.4	0	---
Riparian woodland proper	22	13	59.1	1	7.7
Total Serengeti	86	35	40.7	11	31.4
Ngorongoro	14	2	14.3	2	100.0

Table 4.23: Average persistence of medium-sized adult carcasses in different habitats of the Seronera study area. Carcass samples from which persistence was calculated are those for which complete consumption was observed. The habitat-specific proportion of carcasses observed until complete consumption is not correlated to persistence, indicating a lack of bias in the extensiveness of carcass monitoring among habitats. Using the Mann-Whitney U-test, all paired differences in persistence are significant to the .05 level except for riparian woodland margin versus riparian woodland proper.

Habitat	Total Carcass n	Carcasses observed until completely consumed n	%	Persistence (hours) mean ± S.D.
Acacia woodland	10	8	80.0	10.3 ± 14.9
Plains	54	28	51.9	18.0 ± 8.4
Riparian woodland margin	7	5	71.4	33.4 ± 15.2
Riparian woodland proper	14	7	50.0	70.7 ± 48.8

analysis is restricted to Seronera due to the irregularity of carcass monitoring in other study areas. That data from Seronera are not subject to methodological biases is indicated by the lack of correlation between habitat-specific persistence and the proportion of carcasses observed until consumption was complete (Table 4.23).

Medium-sized adult carcasses persist for almost four times longer in riparian woodland (mean of 71 hours) than in *Acacia* woodland (10 hours) or grass plains (18 hours). Carcass persistence on the margins of riparian woodland (33 hours) is significantly lower than in riparian woodland proper, but still higher than in more open habitats.

The long persistence of carcasses in riparian woodland is in part due to the dominance of lions as initial consumers in this habitat and their tendency to "guard" carcasses between meals until all flesh has been consumed (e.g. Schaller 1972). To test the impact of this possible bias, persistence data for only lion kills in Seronera are presented in Table 4.24. By removing those carcasses very quickly eaten by hyenas as initial consumers, persistence increases slightly for the plains and *Acacia* woodland habitats, but not to a degree significantly affecting the drastic differences seen earlier. Lion kills are therefore abandoned and completely scavenged by hyenas within a day in open vegetation habitats, but on average persist unattended for a far longer interval in riparian woodland.

When hyenas finally did scavenge lion kills in riparian woodland, they left the carcass more complete than in more open habitats (Table 4.25, Figure 4.6). In *Acacia* woodland, hyenas completely scavenged the three lion kills observed. Only some head contents were occasionally ignored by hyenas scavenging lion kills on the margins of riparian woodland. Lion kills were scavenged less thoroughly by hyenas on the plains, but still to a significantly greater extent than in riparian woodland proper. Small scavenging group sizes of hyenas in riparian woodland are apparently responsible for their abandonment of some of the flesh and marrow, and most of the head contents left by lions. Combined with the much longer persistence of carcasses in riparian woodland, this habitat offers by far the best feeding opportunity for non-hyaenid scavengers.

There is an important seasonal component to the scavenging opportunity described for riparian woodland. Lion kills, which account for most deaths in riparian woodland, are most likely to be made in this habitat during the dry season (Table 4.26). Approximately 80% of lion kills within or on the margins of riverine woodland in Seronera occurred during dry days, while most lion kills (75%) in open habitats away from water occurred on wet days. During dry periods, ungulates are forced to drink from perennial water sources within riparian woodland where

Table 4.24: Persistence of medium-sized adult lion kills in different habitats of the Seronera study area. By considering lion kills only, habitat-based differences in persistence are only slightly less than those presented in Table 4.23. Average persistence in riparian woodland proper is significantly greater (.05 level) than that in both *Acacia* woodland and plains (Mann-Whitney U-test, direction predicted).

Habitat	PERSISTENCE (hrs.) n	mean ± S.D.
Acacia woodland	3	18.0 ± 9.1
Plains	14	21.9 ± 12.1
Riparian woodland margin	5	33.4 ± 15.2
Riparian woodland proper	7	70.7 ± 48.8

Table 4.25: Completeness of medium-sized adult lion kills scavenged primarily by hyenas (and occasionally by vultures) in different habitats of the Serengeti. Note the few remains abandoned by scavenging hyenas in the more open habitats and even on the margin of riparian woodland, compared to those abandoned in riparian woodland proper. Differences in means of all habitats paired with riparian woodland proper are significant at least to the .05 level for total carcass and marrow completeness, and for head contents completeness in the cases of *Acacia* woodland and plains (Mann-Whitney U-test, direction predicted).

		MEDIUM-SIZED ADULT PERCENTAGE INDEXED COMPLETENESS: MEAN/(S.D.)				
Habitat	n	Viscera	Flesh	Marrow	Head Conts.	Total
Acacia woodland	3	0	0	0	0	0
Plains	30	0	0.8 (3.1)	6.6 (14.5)	38.6 (39.5)	6.5 (8.4)
Riparian woodland margin	7	0	0	0	35.7 (41.6)	3.6 (4.2)
Riparian woodland proper	16	0	5.0 (15.4)	28.9 (31.1)	59.4 (43.1)	18.8 (19.1)

Figure 4.6: Completeness of medium-sized adult lion kills after scavenging by hyenas in different habitats of the Serengeti. See Table 4.25 for accompanying data.

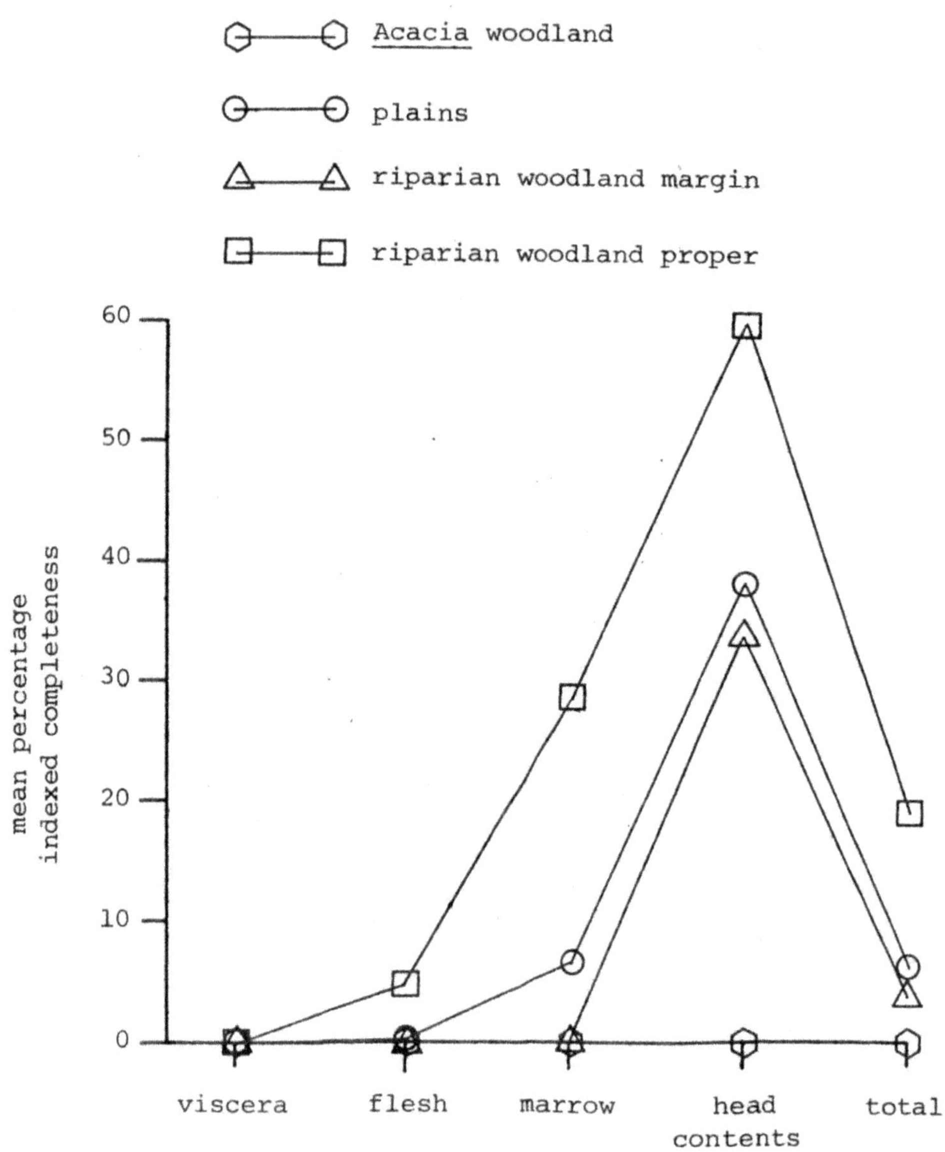

Table 4.26: Distribution of lion kills in the Seronera study area with respect to habitat type and relative amounts of rainfall. Note the greater liklihood of kills to be made in riverine woodland during dry periods, and on the plains during wet periods. $X^2 = 17.12$, d.f. = 1, $p<.001$.

Rainfall	Plains and *Acacia* woodland n	%	Riverine woodland (margin and proper) n	%
Dry	10	25.6	19	79.2
Wet	29	74.4	5	20.8
Total	39	100.0	24	100.0

they are easily ambushed by lions. Wet periods, however, are accompanied by the formation of many temporary pools of water away from permanent sources. During these times, lions are forced to seek prey in open vegetation habitats, where hyenas are plentiful and quick to scavenge most remains. The good scavenging opportunity provided by lion kills in riparian woodlands is therefore primarily a dry season phenomenon.

SUMMARY OF MODERN SCAVENGING OPPORTUNITY

Several clear characteristics of feeding opportunities for scavengers in the Serengeti and Ngorongoro can be condensed from the foregoing analyses. The characteristics can be defined on the basis of the carcass size and age and the associated tissue types most likely to be encountered by scavengers. They can also be defined by the ecological contexts, as conditioned by competition, within which feeding opportunities are the best.

Larger carcasses were shown to persist longer, and, in general, to be abandoned by initial consumers in a more complete state than smaller carcasses. The resultant carcass size dependency of scavenging opportunity incorporates size differences based on age, so that younger individuals of a particular taxon will not persist for as long or in as complete a state as adults.

The contribution of a particular carcass size/age class to a scavenger's diet will also depend on encounter rates. Size-specific encounter rates are conditioned by the composition of the live herbivore community and the differential persistence of carcasses of each size group when being consumed. Previously I suggested the size and age composition of the total carcass sample (Table 2.4) to be a good representation of actual carcass encounter rates by a diurnal scavenger. Medium-sized adults and small adults are the most common herbivores in the Serengeti and Ngorongoro, but their differential persistence as carcasses results in much higher encounter rates with medium-sized adults. The greater weight of tissues remaining on comparably complete carcasses of the two size classes further contributes to the difference in scavenging opportunity provided. Larger herbivores are relatively rare in the Serengeti and Ngorongoro, and their long persistence as carcasses results in encounter rates as a carcass only slightly higher than live encounter rates (Tables 2.2 and 2.3). Although large carcasses will almost always provide a high-yielding meal for scavengers, such bonanzas are very rare. Medium-sized adult carcasses provide the overall best opportunity for scavengers, owing to the greater yield of tissues than provided by small carcasses, and the extreme rarity of larger ones.

The tissue types available to a scavenger are to a large degree dependent on the size and age of carcasses encountered. Larger carcasses generally survive initial consumption with increasingly greater quantities of flesh, while smaller carcasses are likely to provide tissues increasingly representative of the lower-yielding parts comprising the later stages of the consumption sequence. This principle is qualified by carcass age and the number and type of initial consumers. For example, young carcasses are less likely to persist with the brain while adults of the same size are less likely to remain with marrow bones. Carcasses initially fed on by hyenas, especially large numbers of hyenas, are less likely to persist with marrow and head contents than those consumed by smaller numbers of hyenas or carnivores limited in their ability to gain access to within-bone tissues. Medium-sized adults, the carcass class most important to scavengers, will in only special circumstances be abandoned by initial consumers with more than small pockets of flesh on the head, neck and lower limbs. If abandoned with any edible parts, these carcasses much more frequently provide head contents and usually marrow bones. This is invariably the case with medium-sized adults initially consumed by lions and vultures. Adult wildebeest and zebra carcasses initially fed on by hyenas are much less likely to survive with marrow bones or even head contents.

A number of ecological contexts were described among which scavenging pressure by hyenas varies. In most situations (always in Ngorongoro, and in grass plains or *Acacia* woodland of the Serengeti) predatory and scavenging hyenas feed on carcasses in large enough numbers to deprive other scavengers of all but the most marginal opportunity. The only circumstances in which hyenas do abandon substantial remains of medium-sized adults are in areas of the Serengeti temporarily supporting very large concentrations of herbivores during dry periods of high natural mortality. Indeed, it was this context of carcass glut that produced the greatest seasonally-based departures from typical availabilities of scavengeable food.

In another set of contexts defined by habitat type, competition for carcasses by hyenas varied not so much in the completeness by which carcasses are scavenged, but in the amount of time required for them to discover carcasses. In relatively open vegetation habitats away from permanent water sources, hyenas are common and have a very rapid and thorough impact on carcass consumption. Riparian habitats supporting relatively dense woodland are not frequented by hyenas so that lion kills abandoned along rivers and lakes during the dry season persist typically for three days. Vultures are usually present to scavenge immediately most of the small quantity of flesh abandoned by lions, but the full complement of marrow bones and head contents persists undiscovered by hyenas for several days, or until the remains decompose. Such is the consistency of this opportunity that there appears to be a dry season niche available for a scavenger able to expose tissues within bones of medium-sized adult carcasses killed, defleshed and abandoned by lions. The relevance of this and other opportunities for a hominid-like scavenger form the basis for discussion in the following chapter.

CHAPTER 5

EARLY HOMINID SCAVENGING OPPORTUNITIES

The ecology of modern carcass availability has clear and important implications for the scale and characteristics of early hominid scavenging opportunity. In Chapter 4, I documented a number of contexts among which feeding opportunities for modern scavengers vary in expected and regular ways. These observations are directly relevant to the identification of those contexts in which a scavenger with hominid characteristics would be most successful. The accuracy of statements on prehistoric hominid scavenging rest on the premise that the processes of scavenging observed in the present are applicable to the past, being contigent only on the degree to which broad features of prehistoric ecosystem dynamics and hominid adaptation are understood. Testing the implications for the nature of hominid scavenging, and even the existence of scavenging by hominids, is another matter. These tests will require analyses of archaeological bone assemblages directed specifically to hypotheses arising from the implications presented here. Only a brief and preliminary application of results to currently available archaeological evidence will be made here. Most discussion will center on the types of archaeological data relevant to testing both the existence of hominid scavenging and the scavenging opportunities suggested to have been encountered by hominids.

The discussion of hominid scavenging opportunity will be restricted to that in East Africa during the Pliocene-Pleistocene transition, roughly 1.8 to 1.6 million years ago. The earliest evidence for meat-eating by hominids, and, indeed, the earliest archaeological evidence of any kind, is found at the Plio-Pleistocene boundary. The evidence is best seen in a number of minimally disturbed concentrations of stone artifacts and fragmented animal bones recovered from the site complexes of Olduvai Gorge, Tanzania, and East Lake Turkana, Kenya (M.D. Leakey 1971, Isaac and Harris 1978, Isaac in prep., Bunn 1982, Potts 1982). If scavenging ever was an important means of animal food acquisition for hominids, it is most likely to have been so during the earliest phases of the dietary revolution when hominids were least suitably equipped to hunt. Concentrating on prehistoric scavenging opportunities in East Africa has a further advantage. It will be seen that the broad structure of ecosystems in East Africa has not changed radically in the past several million years from that observed in the actualistic field study reported here.

Concentrating on the Plio-Pleistocene of East Africa does not preclude the applicability of modern observations on scavenging ecology to other regions or time periods of prehistory. The results of the actualistic field study are considered to have broad applications to prehistoric lifestyles over the whole time span and many of the regions in which human dietary strategies evolved. In this sense, the patterns documented earlier and their application presented below can serve as an example for other archaeologists with different temporal and regional interests.

PREHISTORIC COMMUNITY STRUCTURE AND DYNAMICS

Applicability of Modern Observations

Neither the modern Serengeti nor Ngorongoro Crater can or will be used as direct, one-to-one analogues to Plio-Pleistocene ecosystems in which hominids foraged. On the one hand, both ecosystems, despite providing a wide range of scavenging opportunities, represent only part of the range of ecological contexts found in East Africa today. For example, neither ecosystem is representative of the dispersal-dominated systems seen in areas such as Amboseli National Park, where herbivores disperse over large areas during wet periods and concentrate around perennial water sources in the dry season (Western 1975). In these terms, the Serengeti and Ngorongoro are atypical of many modern African ecosystems, and to an undetermined degree prehistoric ones as well. The Serengeti is one of the few remaining migratory systems in Africa owing to severe ecosystem truncation by human activities elsewhere (e.g. Houston 1979). Ngorongoro is atypical in being a stable system with respect to annual herbivore numbers, water availability and grazing conditions (see Chapter 2 for a fuller discussion). Additionally, Plio-Pleistocene carnivore and herbivore community structure differs in several important ways from any modern East African community. These differences alone denote important differences between modern and prehistoric scavenging opportunities.

Rather, the worth and design of the observations made in modern settings lies in their clarification of the processes of scavenging. I have identified two major sources of variability in the scale of scavenging opportunity and have shown how these affect the characteristics of the opportunities encountered in two modern ecosystems. Both sources of variability (carcass size in relation to consumer size and number, and the degree of competition for carcasses) are based on fundamental physical and ecological principles. The manner, though not necessarily degree, to which these affect scavenging opportunities today should be similar to their effect in the past.

Hence, large carcasses will always persist longer and be abandoned more complete than small carcasses eaten by the same consumer, virtually regardless of temporal differences in herbivore and carnivore community structure. Only the relative scale of the feeding opportunities provided by small and large carcasses will differ on the basis of possible changes in the size-structure of the herbivore community and the composition of the carnivore community with respect to those able to overcome mechanical constraints to complete carcass consumption. Given a knowledge of past community structure, one should be able to predict the direction of change, if any, in the size classes of carcasses most frequently available to prehistoric scavengers from the size classes seen in the modern ecosystems studied.

Likewise, contexts of intense competition for carcasses will always produce poorer opportunities for scavengers than when competition is low. Again, deviations will be ones of degree based on different scavenger to carcass ratios. In dispersal ecosystems dependent on water availability, competition should be greatest during maximum dispersion of ungulates during the rainy season, much in the same way that wet periods produce high competition in the Serengeti woodlands when all but small numbers of residential ungulates have migrated to the plains. Variations in scavenger to carcass ratios have been shown to be tied to seasonal differences, differences in predator to prey ratios and habitat differences. Given a knowledge of the degree of seasonality and distribution of habitats in past ecosystems, one should be able to predict the contexts, if any, in which competition for carcasses was low and hominid scavenging most successful.

Changes in East African Community Structure

General Statements: Savanna-woodland mosaic ecosystems dominated by medium-sized ungulates have been in existence in East Africa as the dominant ecosystem type for at least the last three million years (e.g. van Couvering 1980). Savanna-woodland mosaics are maintained by a marked seasonality in rainfall (e.g. Sinclair 1975). The existence of seasonal, relatively open vegetation environments three million years ago was the culmination of a global drying trend away from the relatively aseasonal, humid forests typical of the Middle Miocene of East Africa (van Couvering 1980, Andrews and Evans 1979, Andrews et al 1979). After three million years ago, there is evidence for fluctuations in rainfall and the relative extent of different vegetation types (e.g. Bonnefille 1976, Carr 1976, Hay 1976), including several dramatic and partially reversed shifts toward drier conditions associated with decreases in the extent of woodland at 2.5 myr ago and between 2.0 and 1.7

myr ago (Bonnefille 1984, Kappelman 1984). However, since the Late Pliocene, the overall character of the climate, vegetation and fauna have remained similar. Most modern larger carnivore and herbivore genera and many extant species had made their appearance by the Plio-Pleistocene (Maglio 1978, Table 30.1).

Herbivore Community Structure: The taxonomic composition of East African Plio-Pleistocene larger mammalian herbivores is shown in Table 5.1. The data are derived from members G through L of the Shungura Formation on the Lower Omo River of Ethiopia, ranging in age from 1.93 to 1.2 million years ago (Brown and Nash 1976), the Upper Member of the Koobi Formation at East Lake Turkana (< 1.8 million years ago), and Bed I and lower Bed II at Olduvai (< 1.8 mya). Taxonomic data are taken mainly from the comprehensive faunal lists of Coppens and Howell (1976) for the Shungura Formation, and Harris (1978) for the Koobi Fora Formation. Data from Olduvai Gorge are derived from a number of independent sources cited in the caption to Table 5.1. The three localities represent mainly mesic micro-habitats within savanna-woodland mosaic, as determined through palynological, faunal and sedimentological studies (e.g. Bonnefille 1976, 1984, Gentry 1976, Harris 1976, Hay 1976, Kappelman 1984). An inventory of the larger herbivores present historically in East African savanna-woodland environments (Dorst and Dandelot 1969) is also given in the table.

The taxonomic diversity of larger herbivore taxa was much greater in the Plio-Pleistocene than in modern settings (Table 5.1). Most notable are 1) the presence of three elephant-like genera in the Plio-Pleistocene compared to one today, 2) among perrisodactyls, the presence of the last representative (*Ancylotherium*) of the extinct family chalicotheridae, and the genus *Hipparion* in addition to early *Equus*, 3) among suiformes, a plethora of hippopotamids and especially suids of diverse size compared to the persistence of only one species of each today, 4) a great diversity of giraffids including the extinct *Sivatherium*, and 5) among bovids, the presence of *Pelorovis* in addition to the precursor to the modern buffalo (*Syncerus*), and a high diversity of reduncines and alcelaphines, including the extinct giant hartebeest, *Megalotragus*.

The enhanced diversity of Plio-Pleistocene herbivores appears to be real rather than an artifact of methods. Although the Plio-Pleistocene sample spans over half a million years (as in the Shungura Formation), most taxa are present throughout the whole or most of the sequences, such that they represent a synchronous fauna. A greater degree of taxonomic splitting also does not appear to account for the greater diversity of the fossil fauna, as it is taxonomically rich both on the species and genus level (see Table 5.2).

Table 5.1: Larger mammalian herbivores of Plio-Pleistocene and modern East African savanna woodlands. Very small neotragine and cephalophine bovids are excluded. Taxonomic data from the Shungura Formation are from Coppens and Howell (1976). Data from the Upper Member of the Koobi Fora Formation are from Harris (1978). For Olduvai, data on bovids are from Gentry and Gentry (1978), suids from Harris and White (1979), rhinos (Hooijer 1969), *Equus* (Churcher 1982), and the remainder from M.D. Leakey (1971). Size groups are equivalent to those used throughout. An "x" denotes presence, and an "?" an uncertain presence.

Taxon	Size Gp.	Shungura Mem. G-L	Koobi Fora Upper Mem.	Olduvai I + L.II	Modern
Proboscideans					
Deinotherium bozasi	6	x	x	x	
Elephas recki	6	x	x	x	
Loxodonta sp.	6	x	x		x
Perrisodactyla					
Ceratotherium simum	5	x	x	x	x
Diceros bicornis	5	x	x		x
Ancylotherium sp.	5?	x		x	
Hipparion cf. *sitifense*	3	x		x	
H. ethiopicum	3	x	x		
Equus sp. A	3	x	x	x	x
Equus sp. B	3	x	x	x	
Suiformes					
Hippopotamus gorgops	5	x	x	x	
H. protamphibius	5	x	?		
H. aethiopicus	4	?	x		
H. amphibius	5				x
Hexaprotodon karunensis	5		x		
Mesochoerus limnetes	3	x	x	x	
M. majus	2-3	x			
Notochoerus scotti	2-3	x			
Metridiochoerus modestus	2-3	x	x	x	
M. andrewsi	3	x	x	x	
M. hopwoodi	2-3	x			
M. compactus	2	x			
Phacochoerus aethiopicus	2	x		x	x
Potamochoerus sp.	2-3	x			
Giraffidae					
Sivatherium sp.	5	x	x	x	
Giraffa jumae	5	x	x	x	
G. gracilis	4	x	x	x	
G. pygmaeus	4		x	x	
G. camelopardus	5				x
Okapia sp.	3			x	

(continued)

Table 5.1 continued

Taxon	Size Gp.	Shungura Mem. G-L	Koobi Fora Upper Mem.	Olduvai I + L.II	Modern
Camelidae					
Camelus sp.	4-5	x	x		
Tragelaphini					
Tragelaphus nukae	3	x	x		
T. strepsiceros	3	x	x	x	x
T. gaudryi	3?	x			
T. imberbis	3				x
T. scriptus	2				x
Taurotragus sp.	4	x			x
Bovini					
Syncerus sp.	4	x	?		x
Pelorovis sp. A	4-5	x	x	x	
Pelorovis sp. nov. B	4-5	?	x		
Reduncini					
Kobus kob	2	x	x	x	x
K. sigmoidalis	3	x	x	x	
K. ellipsiprymnus	3	x	x	x	x
K. ancystrocera	3	x	x		
Menelikia lyrocera	3?	x	x		
Redunca sp.	2	x	x	x	x
Hippotragini					
Hippotragus gigas	3?	?	x	x	
H. niger	3				x
Oryx sp.	3		x	x	x
Alcelaphini					
Connochaetes sp.	3	x	x	x	x
Megalotragus kattwinkeli	4	x	x	x	
Beatragus antiqus	3?	x		x	
Parmularius angusticornis	3?	?	x	x	
P. altidens	3	x	x	x	
Alcelaphus sp.	3				x
Damaliscus korrigum	3				x
Aepyceros sp.	2	x	x	x	x
Antilopini					
Antidorcas recki	1	x	x	x	
Gazella sp. A	1-2	x	x	x	x
Gazella sp. B	1-2	x	x	x	x
Gazella sp. C	1-2		x	x	

Table 5.2: Size-specific taxonomic diversity of modern and fossil East African larger mammalian herbivores. Data are taken from Table 5.1. Taxa in Table 5.1 given a range of size group estimates that overlap groups used here are placed in the smaller group. Note that the number of fossil taxa are the maximum from one of the three site complexes within the broad taxonomic groups used in Table 5.1. This minimizes any artificially increased diversity resulting from possible synonomy of taxa given different binomens at different sites.

Size Group	SPECIES Fossil n	%	Modern n	%	GENERA Fossil n	%	Modern n	%
Small (Size 1-2)	13	25.5	7	30.4	10	27.0	6	28.6
Medium (Size 3)	20	39.2	9	39.1	13	35.1	8	38.1
Large (Size 4-6)	18	35.3	7	30.4	14	37.8	7	33.3
TOTAL	51	100.0	23	99.9	37	99.9	21	100.0

Table 5.2 compares the size-specific species and generic diversity of the past and present larger herbivore communities of East Africa. Size classes are the same as those described for the carcass sample in Table 2.4. Size class estimates of fossil taxa shown in Table 5.1 are based mostly on those provided by Bunn (1982, Table 2.2) and in a few cases on the size of apparent modern descendents. Size groups presented in Table 5.2 are lumped into three categories owing to uncertainties in the size estimates of some fossil taxa. Hence, small taxa include size 1 and 2 species, medium-sized taxa are size 3, and large taxa lump size groups 4 through 6.

The enhanced taxonomic diversity of Plio-Pleistocene herbivores is shared equally among all three size classes (Table 5.2). Compared to the size distribution of modern herbivores, large taxa account for only a slightly higher proportion of all species (35.3% versus 30.4%) and all genera (37.8% versus 33.3%) in the Plio-Pleistocene. Medium-sized herbivores were equally represented among all species (39.2% versus 39.1%) and slightly less well represented generically (35.1% versus 38.1%) in the Plio-Pleistocene. The relative diversity of small taxa was also slightly less (25.5% versus 30.4% of all species, 27.0% versus 28.6% of all genera). The decreased diversity of herbivore taxa over time apparently occurred to a similar extent in all size classes, so that the size structure of the herbivore community has remained uniform over at least the last two million years.

Carnivore Community Structure: The taxonomic composition of larger mammalian carnivores during the Plio-Pleistocene and the present is shown in Table 5.3. Data for the Plio-Pleistocene are from the same localities and time periods as those from which herbivore diversity was described. Howell and Petter (1976) describe the carnivore finds from the Shungura Formation, while M.G. Leakey (1976) provides the inventory of carnivores from the Koobi Fora Formation, and Petter (1973) that for Bed I and lower Bed II at Olduvai.

As with herbivores, the total diversity of larger carnivores was greater during the Plio-Pleistocene in East Africa than at present (Tables 5.3 and 5.4). Felids, however, account exclusively for the greater diversity. In addition to two species of *Panthera* (lion and leopard) and *Acinonyx* (cheetah) persisting until the present, the Plio-Pleistocene supported three other species of large felids. These include the true saber-tooth cats, *Homotherium* and *Megantereon*, and the "false" saber-tooth, *Dinofelis*. *Homotherium* and *Dinofelis* are lion-sized carnivores, and *Megantereon* is similar in size to the leopard (e.g. M.G. Leakey 1976). The savanna-woodlands of the Plio-Pleistocene supported five genera and six species of larger felids, of which only two genera and three species survive today.

Table 5.3: Larger mammalian carnivores of Plio-Pleistocene and modern East African savanna woodlands. Taxonomic data from the Shungura Formation are from Howell and Petter (1976). Those from the Koobi Fora Formation are from Meave Leakey (1976), while data from Olduvai are from Petter (1973) and F.C. Howell (pers. comm.). "x" denotes presence.

Taxon	Shungura Mem. G-L	Koobi Fora Upper Mem.	Olduvai Beds I + Lower II	Modern
Felidae				
Homotherium sp.	x	x	?[c]	
Megantereon sp.	x	x	x[c]	
Dinofelis sp.	x	x	x[c]	
Panthera leo	x	x	x	x
P. pardus	x		x	x
Acinonyx sp.[a]	x	x	x	x
Hyaenidae				
Hyaena hyaena	x	x	x	x
Crocuta crocuta	x	x	x	x
Canidae				
Canis mesomelas		x	x	x
C. aureus				x
C. adustus				x
Lycaon sp.[b]		x		x

[a] synonomous with *Panthera crassidens* (F.C. Howell, pers. comm.).

[b] synonomous with *Canis africanus* (F.C. Howell, pers. comm.).

[c] Petter (1973) had identified early Olduvai machairodonts as *Machairodus* and *Metailurus*. The former is a Pliocene form from southern Africa (Hendey 1974) and certainly not present at Olduvai, while the latter is probably representative of *Acinonyx* (F.C. Howell, pers. comm.). Howell reports the presence of *Dinofelis* and *Megantereon* [formerly Machairodontinae indet. (e.g. M.D. Leakey 1971)], but is uncertain of the presence of *Homotherium*.

Table 5.4: Taxonomic diversity of modern and fossil larger carnivore taxa in East Africa. Data are taken from Table 5.3.

Carnivore Family	NUMBER OF SPECIES		NUMBER OF GENERA	
	Fossil	Modern	Fossil	Modern
Felidae	6	3	5	2
Hyaenidae	2	2	2	2
Canidae	2	4	2	2
TOTAL	10	9	9	6
TOTAL, excluding canids*	8	5	7	4

* The low fossil diversity of canids is probably related to taphonomic biases.

Members of the diverse felid guild of the Plio-Pleistocene can be suggested to have shown to following prey type specializations. Kurten (1968) suggests at least one of the saber-tooth cats (*Megantereon*) to have been restricted by its relatively short, feeble hindlimbs to large, slow-moving prey such as young proboscideans or rhinoceros. The large canine teeth of both true saber-toothed felids may have permitted a specialization on the high diversity of Plio-Pleistocene large herbivores, by using their canines to stab (*Megantereon*) or slash (*Homotherium*) their prey (cf. Kurten 1968). The Plio-Pleistocene lion along with *Dinofelis* may have concentrated on the size of prey it does today (medium-sized herbivores), the leopard on size 2 herbivores, and the cheetah on smaller prey.

The diversity of hyaenids has remained the same since the Plio-Pleistocene in East Africa, being represented by the modern spotted hyena (*Crocuta crocuta*) and the modern striped hyena (*Hyaena hyaena*). The diversity attained by the Early Pleistocene is in marked contrast to that seen in the preceding Pliocene of both eastern and southern Africa. As many as seven species of a larger size range were present in any one region of the continent during the Pliocene (Ewer 1967, Hendey 1974). The reasons for the extinction of so many hyaenids is unknown (although it may have involved competition with hominids). The two species remaining by the Plio-Pleistocene are similar in size to their modern descendents (Ewer 1967, Petter 1973, F.C. Howell, pers. comm.).

The Plio-Pleistocene fossil record for canids is scant. Only a handful of specimens has been recovered from the Koobi Fora Formation and Olduvai, and as yet none from the Shungura Formation. The wild dog (*Lycaon*) is present in the Plio-Pleistocene, but so far only one of the three extant species of jackal (*Canis mesomelas*) has been recovered. Jackals, however, played a very minor role in reducing scavenging opportunities in the Serengeti and Ngorongoro. The low fossil diversity of canids in general, probably resulting from taphonomic biases, is not deemed significant for my purposes.

Prehistoric Carcass Availability

A general indication of the relative availability of carcasses to scavengers in Plio-Pleistocene and modern East Africa can be given by comparing the relative diversity of larger carnivores and herbivores at each time. In that greater diversity does not necessarily correspond to greater abundance, I do not pretend in the following discussion to be identifying absolute differences in the degree of competition for carcasses during the two time periods. Nonetheless, the data on diversity are suggestive of a possibly important temporal shift in at least the variety if not also the scale of scavenging opportunities in East Africa.

The ratios of larger carnivore to larger herbivore species and genera are shown in Table 5.5 for the Plio-Pleistocene and the present. Among carnivores, only felids and hyaenids are considered in recognition of the minor role canids were observed to play in conditioning modern scavenging opportunities. The ratio of carnivore to herbivore genera is shown to be identical for the two time periods (about one carnivore genus for every five herbivore genera), while the specific ratio is slightly higher during the Plio-Pleistocene. These data are suggestive of an overall similarity in predator-prey relationships between the two periods, with the greater diversity of herbivores during the Plio-Pleistocene being matched by a correspondingly greater diversity of carnivores.

The similarity, however, masks differences in the structure of the carnivore guild in terms of the completeness by which carcasses can be consumed. I am here specifically referring to the fact that the enhanced carnivore diversity of the Plio-Pleistocene is completely accounted for by felids, which are primarily flesh eaters. The extinct saber-tooth and false saber-tooth cats displayed, like all modern felids, large carnassial teeth for shearing flesh but extremely dimunitive premolars. It is the large molars of canids and additionally the enormous premolars of hyaenids that contribute to their ability to crack bones containing nutrients. No felid of the Plio-Pleistocene was larger than the extant lion, implying the inability of all of them, like the modern lion, to gain access to marrow and most head contents of medium-sized adult and larger carcasses. The hyenas were the only common larger carnivore of the Plio-Pleistocene with the potential for completely consuming medium-sized and larger carcasses, including the parts of those abandoned by felids.

Table 5.6 shows the relative diversities of felids to herbivores and hyaenids to herbivores during the Plio-Pleistocene and the present. The comparison is restricted to medium-sized and larger herbivores, as these taxa are large enough to prohibit access by felids to nutrient rich bone contents.

The ratio of exclusively flesh-eating felid taxa to medium-sized and large herbivores in the Plio-Pleistocene is similar to today's ratio (Table 5.6). This is the case on both the generic level and especially the specific level. Major differences, however, are seen between the two time periods in the ratio of more eclectic, bone-cracking hyeanid carnivores to large herbivores. On the species and generic levels, the relative number of bone-cracking carnivore to herbivore taxa in the Plio-Pleistocene exceeds or is nearly equal to two times their current diversity.

Table 5.5: Relative taxonomic diversity of larger herbivores and larger carnivores in Plio-Pleistocene and modern East African settings. Data are taken from Tables 5.2 and 5.4. Canids are excluded owing to their biased underrepresentation in the fossil record, and the minimal role they played in conditioning modern scavenging opportunity.

	NUMBER OF SPECIES		NUMBER OF GENERA	
	Fossil	Modern	Fossil	Modern
Carnivores (felids and hyaenids)	8	5	7	4
Herbivores	51	23	37	21
Ratio of carnivores to herbivores	1:6.4	1:4.6	1:5.3	1:5.3

Table 5.6: Relative taxonomic diversity of modern and fossil carnivores and herbivores. Only medium-sized and larger herbivores are considered. Carnivores are separated according to those able to extract within-bone tissues on medium-sized and larger carcasses (hyaenids) and exclusive flesh eaters (felids).

	NUMBER OF SPECIES		NUMBER OF GENERA	
	Fossil	Modern	Fossil	Modern
Exclusive flesh eaters (felids)	6	3	5	2
Flesh eaters and bone crackers (hyaenids)	2	2	2	2
Medium-sized and larger herbivores	38	16	27	15
Ratio flesh eaters to larger herbivores	1:6.3	1:5.3	1:5.4	1:7.5
Ration bone crackers to larger herbivores	1:19.0	1:8.0	1:13.5	1:7.5

In the Plio-Pleistocene, therefore, there existed a greater diversity of exclusively flesh-eating predators that preyed on a greater diversity of herbivores. That a similar degree of diversity is not seen among Plio-Pleistocene hyaenids suggests two alternatives: 1) hyeanids of the Plio-Pleistocene were much more abundant than today, being able to utilize fully the carcass parts remaining from a greater size diversity of felid kills, or 2) relative to the present, there existed in the Plio-Pleistocene a more frequent availability of marrow bones and head contents, much of which went unscavenged by hyenas. In the absence of data on the relative *abundance* of herbivores, hyaenids and felids in the Plio-Pleistocene, the accuracy of each alternative cannot be tested. The available data on taxonomic diversity suggest, however, that if opportunities for a non-hyaenid scavenger differ between the Plio-Pleistocene and the present, the difference is more likely to be in the direction of a *better* opportunity during the Plio-Pleistocene, and in particular for a scavenger able to extract within-bone tissues of medium- and large-sized carcasses abandoned by felids.

THE HOMINID SCAVENGING NICHE

Relevant Hominid Characteristics

There are several well-documented features of Plio-Pleistocene hominid adaptation relevant to the types of scavenging opportunity hominids might have encountered. These features all pertain to the hominid species responsible for the formation of archaeological sites bearing evidence for meat-eating. Of the two hominid species existing during the Plio-Pleistocene of East Africa, *Homo habilis* and *Australopithecus boisei*, the former is usually considered to be the exclusive creator of the archaeological record.

An important adaptive feature of early *Homo* is their use of stone tools to slice meat from bone and extract marrow from limb bones (Bunn 1981, 1982, Potts and Shipman 1981). The array of informal stone tools used by early hominids includes sharp-edged flakes of stone. These have been shown experimentally to produce the same types of fine linear striae in the process of defleshing and disarticulating bone that are found on many fossil specimens from Plio-Pleistocene archaeological sites. The stone tool kit also contains naturally or artificially rounded pounding stones which experimentally produce the type of impact notches on bone when cracked for its marrow that are also seen at the early archaeological sites. These artificial extensions of the body are commonly viewed as essential components of hominid meat-eating, especially from large animals.

Another feature of hominid adaptation relevant to the scavenging opportunities they potentially encountered is their apparent preference for mesic microhabitats. All known early archaeological sites are located in or on the margins of ephemeral or perennial water sources. Plio-Pleistocene archaeological sites of the Lower Omo basin are all in floodplain, levee or deltaic deposits of the paleo-Omo River (e.g. Merrick and Merrick 1976). Those at East Lake Turkana and Olduvai are on the margins or along feeder streams of paleo lakes Olduvai and Turkana (e.g. Hay 1976, Isaac and Harris 1978). These sites are likely to have been within riparian woodland denser than the vegetation cover away from the water sources, as evidenced by palynological and faunal data (e.g. Bonnefille 1984, Kappelman 1984). That sites have not been found outside of mesic microhabitats is a result of preservational biases and not necessarily an exclusive preference by hominids for riparian habitats. Nonetheless, the known sites can best be interpreted as foci of hominid activity that included carcass processing. The question being addressed here is how these carcasses found in riparian settings were acquired.

Two other known features of hominid adaptation are relevant to their potential as scavengers. One is a relatively slow but efficient locomotor system permitting long distance travel (e.g. Rodman and McHenry 1980). It also can be safely assumed that, like other primates (e.g. Wrangham 1977, Milton 1981), hominids were proficient, at least on an individual basis, at monitoring fluctuations in the availability of food, including highly ephemeral carcass resources.

A major uncertainty of hominid adaptation is their rank on the prehistoric carnivore hierarchy. Hominids are noted for their lack of fleetness and physical prowess, which in itself has been suggested to have limited them to scavenging for obtaining food from large carcasses (see Chapter 1). But scavenging success alone can be expected to vary with the extent to which hominids were able to usurp carcasses from larger mammalian carnivores. Binford (1983) has stated as absurd the notion that the small hominids of the Plio-Pleistocene could steal carcasses from large felids or hyaenids. On the other hand, the demonstrated proficiency of chimpanzees at using props (branches most often) to enhance the intensity of their threats in general (e.g. van Lawick-Goodall 1968) and toward carnivores in particular (Kortlandt 1980) suggests that a sufficiently determined and better equipped early hominid may have been a successful competitor for carcasses yet to be abandoned by large carnivores. In the absence of a means for determining the relative subordinance of hominids to other carnivores, I will assume their access to carcasses to have been restricted to those unattended by large felid and hyaenid carnivores. This will minimize the scale of scavenging opportunity assigned to early hominids, but make it consistent with data presented herein on carcass completeness, which measures completeness as abandoned by other consumers.

Potential Contexts of Hominid Scavenging

Table 5.7 reviews the contexts among which scavenging opportunity has been found to vary in the modern Serengeti and Ngorongoro. Three sets of contexts are shown, including carcass size, consumer type, and ecology as affected by seasonality and habitat type. The specific contexts within each set have been rated according to the quality of opportunity provided for a scavenger with the set of hominid characteristics described above. The ratings are determined by qualitatively weighing a number of factors documented earlier to condition scavenging opportunity, including 1) the relative abundance of deaths in a particular context, 2) the relative completeness and tissue composition of carcasses remaining after initial consumption, 3) the relative persistence of carcasses as food resources, and 4) the degree of competition for carcasses, and the effect of competition on reducing the completeness of partially consumed carcasses.

Medium-sized adult and larger carcasses are most likely to have provided the most regular and substantial scavenging opportunity to hominids. As shown in detail in Chapters 3 and 4, smaller animals, including young of larger species, die in large numbers, but they are usually completely eaten by most initial consumers, and, if surviving initial consumption, persist for a very short time until being completely scavenged. Medium-sized adults also die in abundance, and in many cases survive initial consumption with at least marrow and head contents. These remains persist longer than on small carcasses, but when scavenged, are subject to complete consumption in many cases. Large adult carcasses are often abandoned by initial and subsequent consumers with most marrow bones and head contents intact or nearly so, and in the case of very large carcasses, with substantial quantities of flesh as well. These remains will often persist on very large carcasses until decomposition sets in and prohibits further consumption by vertebrate carnivores. The quality of scavenging opportunity provided by large carcasses is very good, except that they are, relative to medium-sized adults, very rare finds. The greater diversity of large herbivores during the Plio-Pleistocene, however, implies their somewhat more frequent occurrence as scavengeable carcasses then.

Hominids equipped with stone tools would have been minimally constrained by the mechanics of large carcasses, both in terms of exposing flesh beneath thick-skinned areas of the body and in exposing marrow and head contents. The use of stone tools to process carcasses, in fact, seems to

Table 5.7: The quality of early hominid scavenging opportunity in different contexts.

Context of Opportunity	Carcass Type(s)	Abundance	Completeness After Initial Consumption	Intensity +/or Effect of Competition	Persistence	Quality of Scavenging Opportunity
CARCASS SIZE						
Small	small adults and subadults	high	low	mostly complete consumption	short	POOR
Medium	mostly medium-sized adults	high	moderate	usually complete consumption	moderate	GOOD
Large	large and very large adults	low	high	many parts putrify	long	GOOD BUT RARE
INITIAL CONSUMER TYPE						
Felids	all	high	low on small prey, high on large prey	complete if hyenas, only flesh if vultures	long	GOOD
Vultures	all	low, only natural deaths	high, but little flesh	size-dependent	long	RARE
Hyenas	all	high	low	rarely any remains to scavenge	short	POOR

(continued)

Table 5.7 continued

Context of Opportunity	Carcass Type(s)	Abundance	Completeness After Initial Consumption	Intensity +/or Effect of Competition	Persistence	Quality of Scavenging Opportunity
ECOLOGY						
High predator:prey ratio (Ngorongoro), or scavenger:carcass ratio (Serengeti: wet, herbivore scarcity, open habitats)	all	low	low	intense competition, complete carcass consumption	short	POOR
Low scavenger:carcass ratios						
-herbivore abundance and dry conditions	mostly abundant species	very high	very high	minimal, mostly from vultures	long	GOOD
-riparian woodland						
-wet periods	---	low	low	low	long	RARE
-dry periods	felid kills	moderate	moderate to high	delayed, except by vultures	long	GOOD

have afforded early hominids the opportunity to utilize all carcass parts, including the within-bone tissues of medium-sized and larger carcasses abandoned by felids. Stone tool use may therefore have given hominids a competitive edge over vultures in exposing flesh well protected by skin in larger carcasses, and over hyenas in exposing tissues within bones of medium-sized and larger carcasses. The advent of stone tool use in carcass processing, in this sense, is seen as a critical feature in the origin of meat- and marrow-eating among hominids. That the timing of the origin of tool use coincides broadly with the marked reduction of hyaenid diversity from the Late Pliocene high to the current low suggests it to be the novel adaptation enabling hominids to enter the apparently recently underexploited scavenging niche. It may even prove correct that stone tool use enabled hominids to exclude many hyaenids from a scavenging niche, and to be at least partly responsible for the extinction of many Pliocene hyaenids.

Within the context of consumer type, carcasses initially consumed by felids are shown to provide the best opportunity for a hominid scavenger (Table 5.7). Felid kills are common today, and perhaps would have been more common and of a larger, more equable size distribution in the past. This results from the greater diversity of Plio-Pleistocene felids, and their suggested individual prey type specializations within the whole range of larger herbivores. It has been suggested on the basis of jaw structure that all felids of the Plio-Pleistocene would have been unable to consume marrow and head contents (save nasal pulp) of carcasses at least the same size as or larger than medium-sized adults. Carcasses consumed by felids would have a long average persistence, resulting in feeding opportunities for a hominid scavenger of all marrow, most head contents, and, on very large carcasses, substantial amounts of flesh.

Vultures and hyenas would provide poorer opportunities to a hominid scavenger. As documented earlier, vultures cannot expose tissues within bones, but except on the largest carcasses are very effective flesh eaters. Vultures are rarely initial consumers, and when they are, it is on natural deaths likely to have very poor quality marrow. The tendency for hyenas to feed in large groups and their ability to consume all parts of even the largest carcasses make them poor candidates for providers of scavengeable food for hominids.

The good quality opportunities for a hominid scavenger provided by larger carcasses and by felid kills can be viewed within contexts of variably intense competition as applied by hyenas. The scale of hominid scavenging opportunity would be poor in contexts of high hyena to carcass ratios, as represented by Ngorongoro Crater at all times of the year, and within the Serengeti during times of low migratory prey abundance, high rainfall, and usually in open vegetation habitats away from water sources. In dispersal systems, low prey abundance corresponds to the rainy season. Dry periods, however, would witness the aggregation of at least water-dependent species (including hominids) within the area of permanent water sources where carcass densities would be relatively high, and competition for carcasses low. Very dry periods of prey adundance in all ecosystems, as indicated in Table 5.7, provide a good opportunity for a hominid-like scavenger. In these circumstances, carcass abundance approaches a point where local scavenger populations are saturated with food, and carcasses persist for relatively long periods with large quantities of flesh. It is only at these times that open vegetation habitats away from rivers and lakes are considered to provide a good opportunity for a hominid scavenger. Wetter periods, with lower concentrations of herbivores and lower rates of natural mortality, provide fewer carcasses and result in the more rapid and thorough consumption of carcasses typical of hyenas in open vegetation habitats.

Aside from high dry season mortality, the other ecological context providing a good opportunity for a hominid-like scavenger is from felid kills in riparian woodland. This setting has been shown to afford long carcass persistence in the absence, or at least extreme slowness, of hyena scavenging. Most felid kills in riparian woodland occur during the dry season when herbivores drink from well-vegetated perennial water sources providing numerous ambush points for felids. Like modern felids, the saber tooth and false saber tooth cats of the Plio-Pleistocene probably relied on stealth and ambush to take down prey after a relatively short chase. The extent to which extinct felids were open plains forms like the cheetah, or, whenever possible took advantage of denser vegetation for ambushing, is not known. The weak-limbed <u>Megantereon</u>, however, probably took advantage of vegetation cover to surprise prey. It can be suggested that riparian woodlands were the favored hunting grounds of most Plio-Pleistocene felids during the dry season. Hyenas, as today, might have ventured infrequently into these habitats to scavenge. At any rate, the evidence for hominids centering activities in riparian woodland suggests they would have had quick access to any felid kills abandoned there.

In summary, two opportunities emerge as the most likely to have been available to a hominid scavenger. One is from abandoned felid kills of medium-sized and larger prey in riparian woodland during the dry season. The other is encountered during periods of prey abundance when additionally natural mortality is high. Both opportunities are primarily dry season phenomena, with comparable and other opportunities during the wet season not being nearly as regular or profitable. The two opportunities, however, differ in a number of respects that have important implications for the degree to which any one was encountered and taken advantage of by hominids. These differences are outlined in Table 5.8 and their implications for hominid scavenging are discussed below.

A major difference between the two opportunities is the habitat type in which the opportunity occurs. In my experience in the Serengeti, virtually all stress-related natural mortality occurred outside of riparian woodland in open vegetation habitats. Dry conditions are stressful to herbivores not so much because of thirst, but because of the increasingly low quality forage they consume during prolonged dry periods (Sinclair 1979a). Natural mortality among grazers such as the alcelaphines, antelopines and equids that dominate the larger herbivore communities of savanna-woodlands, is most likely to occur in their preferred, open vegetation habitats away from riparian habitats. This has also been observed in Nairobi National Park during a severe drought in the early 1970's by Hillman and Hillman (1977). Natural deaths among browsers (e.g. giraffe) and especially herbivores preferring riparian habitats (e.g. elephant, hippopotamus, buffalo, waterbuck) might on the other hand be expected to be concentrated in riparian habitats. This has been demonstrated for elephants during a prolonged drought in the Tsavo National Park by Corfield (1973). As well, the natural deaths of one giraffe and one elephant I observed during my study both occurred in the dense _Acacia_ woodlands around Lake Ndutu. Mesophyllic species, however, occur in far smaller live numbers than xerophyllic grazers in savanna woodland ecosystems. As a result, scavenging opportunities from stress-related natural deaths will be most frequently realized in non-riparian, open vegetation habitats. A hominid scavenger that centers activities in riparian habitats would therefore incur lower travel and search costs when foraging for abandoned felid kills in riparian woodland than it would by foraging into open habitats away from water for natural deaths (Table 5.8). As well, the risk of exposure to predators would be higher in open vegetation habitats than in riparian woodland which provides many trees for easy escape.

While search and risk costs may be lower in riparian woodland, the benefit realized from scavenging abandoned lion kills is also lower than that expected from very complete natural deaths. Abandoned felid kills usually provide only limited quantities of flesh in addition to marrow bones and head contents. Natural deaths, conversely, usually provide a bonanza of flesh, especially if occurring in large numbers. For example, on a very productive day of carcass search in the Simba Kopjes study area on the plains, I discovered six carcasses of adult wildebeest and zebra in an approximately 40 km^2 area. All of the carcasses remained with at least marrow bones and head contents, and three were either abandoned to vultures or discovered

Table 5.8: Characteristics of the two best scavenging opportunities potentially encountered by hominids.

Characteristic	Riparian Woodland Felid Kills	Natural Deaths In Open Vegetation Habitats
Length of foraging trip	short, within areas of sites	long, into habitats away from sites
Risk of exposure to predation	low, many trees for escape	high, few opportunities for escape
Scavengeable tissue yield and type	low, mostly marrow and head contents	high, large amounts of flesh
Nutritional content	high fat, low protein	high protein, no fat
Duration of opportunity	long, throughout all dry periods	short, at end of prolonged dry periods

having been fed on only by vultures, remaining with substantial amounts of flesh on at least the anterior half of the body. A similar magnitude of feeding opportunity was observed on the Naabi-Ndutu plains during several weeks of unusually dry weather when zebra especially were dying in large numbers.

There are important nutritional differences in the parts comprising each of the two scavenging oportunities. Marrow bones of animals killed when otherwise healthy are a good source of calorically rich fat. On only 7 of 27 medium-sized adult lion kills in riparian woodland during the dry season was marrow observed to be low in fat. On the remaining carcasses I found the marrow on visual inspection (Sinclair and Duncan 1972) to be high in fat, or could safely infer their fattiness because they were scavenged by hyenas [on many other occasions, hyenas were observed to ignore marrow bones of stress-related deaths, often after only licking or puncturing the cancellous epiphysis of the bone]. I have determined the twelve major limb bones of a healthy adult male wildebeest, common prey of lions, to provide approximately 350 grams of fat. This amount is equivalent to over 3000 calories (9 cal/gram of fat), which is well over the daily caloric requirements of a small adult modern human (e.g. Lee 1968). Stress-related natural deaths, on the other hand, provide much flesh high in protein but devoid of fat. Wild ungulates in general have very low intramuscular fat levels (usually between 2% and 5% by weight, Ledger 1968), and the early use of intramuscular fat among all body fat deposits as an energy reserve by stressed animals (e.g. Harris 1945, Sinclair and Duncan 1972, Smith 1970) would completely remove all fat from flesh in natural deaths. When nutritionally stressed, as is likely to occur during the dry season, fat contained in marrow bones is the last energy reserve to be used by the animal. If stress continues, all marrow fat is depleted and the animal soon dies. Stress-related natural deaths would therefore provide fat neither from flesh nor limb bone marrow cavities.

Speth and Speilman (1983) describe the nutritional complications, including starvation, that arise from an exclusive protein diet. Even a high protein diet, to be nutritionally sound, must be complemented by large amounts of carbohydrates or fats to help metabolize the proteins. The potential severity of the problem, in fact, has led Speth (in prep.) to argue that flesh-eating by hominids during the dry season would be an unviable nutritional strategy. This would certainly be true if hominids attempted to subsist exclusively on the flesh of stress-related natural deaths for a prolonged period. The extent to which hominids consumed carbohydrates and plant fats, however, is not known. Only recently have two studies been conducted on the seasonal and habitat-based availability of such plant food nutrients (Sept 1984, Vincent 1985). What is certain, however, is that foraging for abandoned lion kills of medium-sized adult and larger carcasses in riparian woodland would provide a concentrated source of calories, and one perhaps large enough to permit the consumption of large amounts of no-fat flesh from natural deaths.

A final important difference between the dry season scavenging opportunities encountered within and outside of riparian woodlands is the frequency and duration over which the opportunities exist (Table 5.8). Periods of high natural mortality are usually restricted to the final two months of the dry season (Houston 1979). As well, for natural deaths to saturate other scavengers, large numbers of herbivores must be in the area in which conditions are stressful. Herbivores, however, are forced to drink from permanent water sources not only throughout the dry season, but also during predominantly wet periods that experience a week or more of no rain. Abandoned felid kills in riparian woodland should be available for at least six months of every year in a region that receives similar amounts of rainfall as the Serengeti. The actual number of abandoned lion kills available to a scavenger in riparian woodland depends on the number of lions present. Along a 10 kilometer stretch of the Seronera River regularly monitored during my study, two lion prides were present and hunting throughout dry periods. From each pride, I learned to expect to find at least one kill every two or three days. Occasionally, multiple kills would be made. One could therefore expect to find at least one abandoned lion kill per day per ten kilometer stretch of the Seronera. As well, the kills were very predictably located. The movements of a pride are very conspicuous, as they would have been to an early hominid group, and kills tend to be localized for a period of time along a short stretch (approximately one kilometer) of riparian woodland. The presence of carcasses could be easily monitored by an early hominid group, and quickly scavenged upon abandonment by lions. That there is no scavenger today specializing on the within-bone remains of lion kills in riparian woodland suggests the availability of a scavenging niche that might have been filled in the past by hominids supplementing a mostly plant food diet with high energy fats from marrow bones.

Foraging Tactics for a Hominid Scavenger

The characteristics of the two good scavenging opportunities described above can be viewed as component costs and benefits of alternative foraging strategies used by hominid scavengers. The purpose of viewing the two opportunities in terms of costs and benefits is to assess their relative viability and to identify the foraging pattern most likely associated with each.

With respect to the foraging pattern used, I am particularly interested in modeling the circumstances under which a hominid scavenger would have foraged most profitably in a "feed-as-you-go" (Isaac 1980, 1981) or "routed" (Binford 1984) mode typical of nonhuman primates, or a central-place foraging mode unique among primates to hunter-gatherers. The type of foraging strategy practiced by early hominids has polarized views among paleoanthropologists.

The central place or home base foraging strategy is represented most explicitly by Isaac (e.g. 1978a, 1978b). Here, early hominid groups are seen to forage from and return to on a daily basis a fixed point on the landscape where gathered foods are distributed among group members in a way similar to that seen in many hunting and gathering groups today. The fixed point, or central place, is seen archaeologically as the many dense concentrations of animal bone and stone tools referred to as sites (or more appropriately "mega-sites", Isaac et al. 1981). Such concentrations clearly could not have been produced in a day, but rather can and have been explained as representing the debris left over repeated daily visits to the locale, as might be expected from a central-place mode of foraging. This view has been supported by other researchers including Bunn (1982). Isaac (1980, 1981, 1983) has more recently suggested a number of other behavioral tactics that could result in such dense concentrations of artifacts and food refuse without invoking the classical home base mode of subsistence.

The feed-as-you-go foraging mode first suggested by Isaac as an alternative to central-place foraging by early hominids, has recently been relabeled as routed foraging and championed by Binford (1984). Routed foraging is less organized than central-place foraging. It involves traveling among a number of resource points on the landscape each day, and remaining at or near the last resource point at the end of the day. No effort is expended to return to the starting point, which might not be revisited for long periods. Routed foraging is typical of all nonhuman primates not tied down to nesting sites (prosimians) or sleeping sites (hamadryas baboons), and implies for early hominids a similar lack of cooperation in foraging and an absence of food sharing among group members. The routed foraging position explains the formation of archaeological mega-sites as having occurred periodically over a relatively long period of time, and representing activities conducted specifically at the resource point being exploited. Aside from its proximity to the resource point (raw material source for stone tool making, fruiting trees, etc.), the sites have no special significance as social foci at which hominid group members assemble on a regular and planned basis.

The type of foraging mode assigned to early hominids clearly corresponds to the general degree of humanness granted to them. Central place versus routed foraging is one more ingredient, similar to the hunting versus scavenging issue, of the differing views on the timing of the origins of unique human traits discussed in Chapter 1. It therefore becomes appropriate and important to evaluate the foraging mode most likely to be associated with a scavenging mode of animal food acquisition. This can be done through a cost-benefit analysis.

The cost-benefit analysis necessarily views hominids as exclusive scavengers. That is, they are purposefully foraging for and consuming only animals discovered dead in the two contexts of good scavenging opportunity. I will for the sake of simplicity and clarity make no attempt to integrate into the model the relative costs and benefits of foraging for plant foods. Hence, the model loses realism (in the sense of Maynard Smith 1974) - hominids certainly have always been omnivores deriving most nutrients from plant foods - but gains specificity by being able to integrate into strategies many of the relevant characteristics of the two good scavenging opportunities potentially encountered by hominids.

Figure 5.1 depicts a model of the costs and benefits of scavenging within and outside of riparian woodland as a function of foraging distance. The origin of the graph represents a point in riparian woodland (e.g. sleeping trees) from which a hypothetical early hominid group commences foraging for carcasses on a given day. Two straight line cost curves (C_a and C_b) are shown. Curve C_a represents exclusively the cost of travel incurred while searching for abandoned felid kills in a routed foraging mode within riparian woodland along a course more or less parallel to a river channel or lake shore. The curve attains its highest point at distance D_{max}, which represents the maximum possible foraging distance that can be covered in a day. The other cost curve, C_b, has a slope twice that of C_a to account for the higher costs associated with several circumstances, including 1) routed foraging for natural deaths outside of riparian habitats where a risk cost is added to account for the scarcity of trees from which to escape predation, 2) central-place foraging in non-riparian, open vegetation habitats for natural deaths, where return trip travel to the central place essentially doubles travel costs (Orians and Pearson 1979, Hamilton and Watt 1970) but reduces risk of exposure to predation by minimizing time spent in areas providing few means of escape from predators, and 3) central-place foraging for abandoned felid kills in riparian woodland, which again is twice as costly as routed feeding owing to return trip travel costs.

The model does not incorporate costs associated with carcass processing as these are assumed to be similar for carcasses found in both habitats. The cost of transporting carcass parts to a central place are included in the additional cost associated with return trip travel.

Figure 5.1: Cost-benefit model of scavenging in riparian or open vegetation habitats as a function of foraging distance.

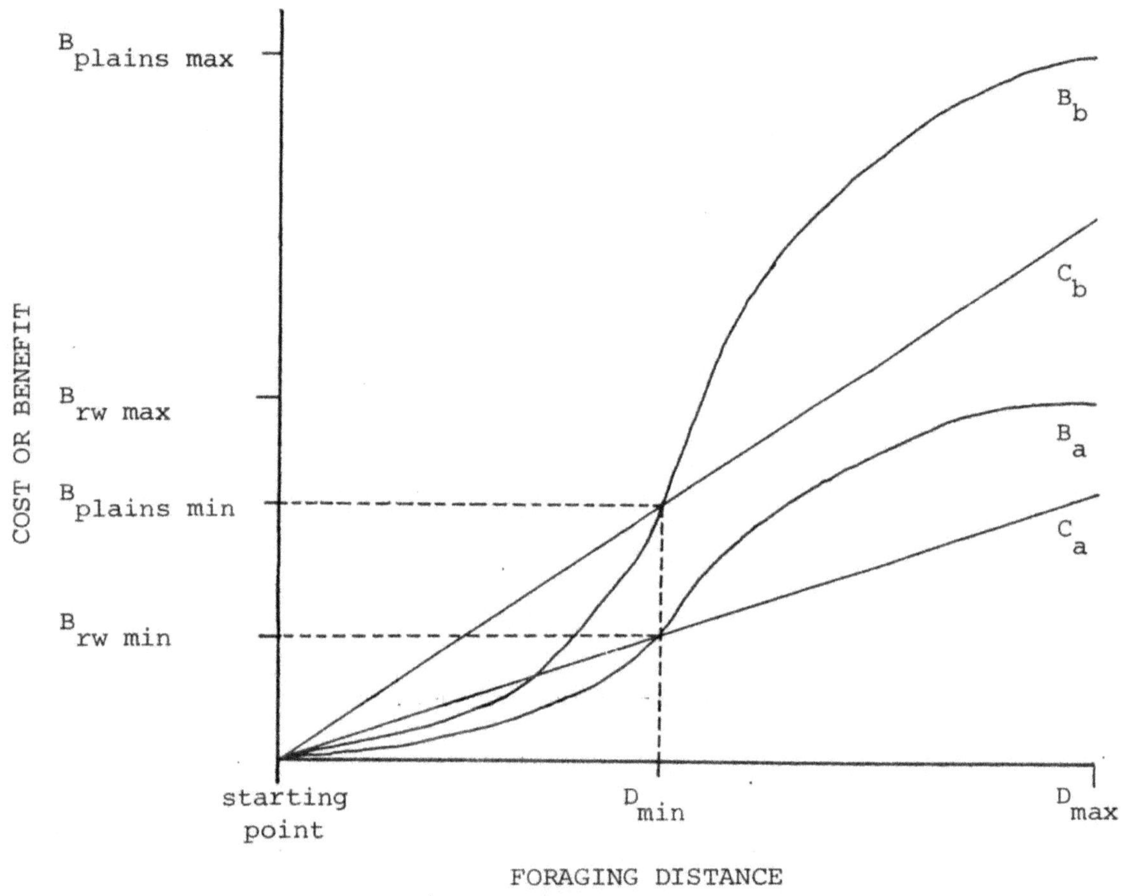

B_a = benefit accrued from scavenging abandoned felid kills in riparian woodland.

B_b = benefit accrued from scavenging natural deaths in open vegetation habitats away from water sources.

C_a = cost of routed foraging in riparian woodland.

C_b = cost of central place foraging, or routed foraging in non-riparian habitats with added risk costs.

D_{min} is the minimum foraging distance at which costs are balanced by benefits accrued in riparian woodland habitats ($B_{rw\ min}$) and open vegetation habitats ($B_{plains\ min}$). D_{max} is the maximum daily foraging distance corresponding to the maximum benefit accruable in riparian woodland ($B_{rw\ max}$) and plains ($B_{plains\ max}$).

Two benefit curves, B_a and B_b, are shown. Their positive slope reflects the increasing yield accrued with increasing foraging distance. Their sigmoid shape reflects the low chance of finding a carcass at the beginning of a foraging trip, and the increased chances of finding a particularly high-yielding carcass as foraging progresses from the starting point. Benefit curve B_a refers solely to the benefits accrued by consuming tissues of abandoned lion kills discovered in riparian woodland. This curve attains its highest point and levels out at a point on the y-axis ($B_{rw\ max}$) corresponding to the maximum benefit available within a day's maximum travel (D_{max}) within riparian woodland. Benefit curve B_b represents the benefit realized by foraging for carcasses in open vegetation habitats during dry periods of prey abundance and high natural mortality. Its slope is shown to be two times that of B_a to reflect the minimum extra yield provided by natural deaths bearing flesh. Curve B_b reaches its highest point at $B_{plains\ max}$ on the y-axis, corresponding to maximum daily foraging distance (D_{max}). $B_{plains\ max}$ is much closer to, or exceeds, the group satiation point than is $B_{rw\ max}$.

The two benefit curves are drawn so as to intersect the corresponding cost curves at the same foraging distance. This has been done to simplify the predictions arising from the model. The points of intersection correspond to foraging distance D_{min} on the x-axis. This point represents the distance at which benefits accrued balance the costs of foraging, and hence represents the minimum distance required to be covered by an exclusive scavenger in each habitat to make scavenging profitable. At shorter foraging distances, scavenging in each habitat is unprofitable (costs outweigh possible benefits) and should not be practiced. The maximum profit, or net benefit, is realized in both habitats at a foraging distance (not shown) between D_{min} and D_{max} where the corresponding benefit and cost curves are most disparate.

Two implications for hominid foraging arise from the model as constructed. First, for scavenging natural deaths in open vegetation habitats to be as profitable as scavenging abandoned lion kills in riparian woodland, the minimum benefit accrued must be twice as high as that accrued within riparian woodlands ($B_{plains\ min}$ versus $B_{rw\ min}$ on the y-axis). This is due to the higher costs derived from additional risk or return trip travel not encountered with routed foraging in riparian woodland. If hominids were tied to a central place <u>and</u> if risk costs were greater than shown, costs of foraging in open vegetation habitats away from water might be so high as to negate a profit and to prohibit scavenging outside of riparian woodland. High costs and low benefits were, in fact, the basis for suggesting that hominid scavenging in hyena-frequented habitats during times of herbivore scarcity and low natural mortality would be unprofitable. Here, however, it is exclusively the high costs incurred that would render the opportunity provided by natural deaths poor and the central-place foraging strategy unprofitable. Whether or not this is the case might depend to a great extent on the degree of group cooperation in anti-predator defense exhibited by early hominids. If effective, such defense could minimize the risk associated with foraging in open habitats, and make scavenging for natural deaths by a central-place foraging mode profitable.

The second implication arising from the model is that anything that inceases foraging costs in riparian woodland will negate any benefits realized from the relatively low yielding lion kills found there. For instance, foraging from a central place in riparian woodland would incur costs described by curve C_b, which at all foraging distances is higher than the benefit curve for riparian habitats. It may well be that foraging for abandonded lion kills in riparian habitats would be insufficiently beneficial to finance a central-place foraging mode there. Rather, the carcasses could be most efficiently scavenged by practicing a routed feeding mode typical of nonhuman primates. In more realistic terms, scavenging abandoned lion kills in riparian habitats would best have been an opportunistic pursuit, with any carcasses being discovered in the course of foraging for plant foods being taken for their calorically rich marrow bones and brain.

In summary, the cost-benefit model suggests scavenging in riparian woodland to be best conducted through a routed foraging mode. The relatively low yield of the remains from abandoned lion kills discovered is suggested to be insufficient to compensate for additional transport and return trip travel costs to a central place. Scavenging for natural deaths outside of riparian woodland, on the other hand, was argued to be profitable only if conducted through central-place foraging which additionally features a high degree of group cooperation in anti-predator defense. Hence, despite the greater yield of scavengeable food available from natural deaths outside of riparian woodland, acquiring these is suggested to be a more costly activity in terms of risk and energy expended than scavenging in riparian woodland.

If early hominids did scavenge, the most readily accessible source acquired in a manner departing the least from the foraging patterns of nonhuman primates, and presumably that of pre-stone tool using hominids, would be that provided by abandoned felid kills in riparian woodland. This source as well would subject hominids to the lowest levels of competition with other scavengers. It would provide mostly medium-sized adult carcasses, occasionally larger ones abandonded by felids, and more rarely carcasses of large mesophyllic animals found dead from natural causes. The opportunity for a scavenger in riparian woodland would be encountered primarily during the dry season as a result of the highly seasonal pattern of felid predation and natural mortality within riparian woodland. The opportunity would provide most regular access to marrow bones and head contents, with flesh being available infrequently and in very small pockets except on the rare large carcass found. Finally, it is precisely in these riparian settings that the greatest diversity and abundance of plant foods potentially eaten by hominids occurs (Sept 1984).

The Archaeological Recognition of Scavenging

The characteristics of hominid scavenging opportunity suggested above have important implications for the archaeological recognition of scavenging. Archaeological faunal assemblages accumulated through scavenging should have the following characteristics:

1) a predominance of adults over juveniles or younger individuals, with most young being those of large species.
2) a predominance of medium-sized adult and larger carcasses over those of smaller size classes.
3) a predominance of fragmented limb and cranial parts over elements from the axial region of the skeleton.
4) a high proportion of all defleshing cut marks occurring on head and lower limb bones, with increasingly lower frequencies being found on parts defleshed progressively earlier in the consumption sequence.
5) smaller taxa being represented by a greater predominance of latterly consumed skeletal parts, and displaying defleshing cut marks on a more limited series of body parts than larger taxa.

Characteristic # 1 above is consistent with the suggestions of Vrba (1975, 1980) and Klein (1982), but for reasons different from those presented here. Vrba cites different frequencies of juveniles over adult animals in the diet of hyenas when scavenging versus when hunting to support a contention that juveniles should be underrepresented in assemblages accumulated primarily through scavenging. Klein, focussing on the apparent association between hunting and catastrophic mortality profiles (as results in game drives) on the one hand, and scavenging and attritional mortality profiles on the other, also states that young individuals should be poorly represented in assemblages accumulated through scavenging. The basis for the distinct age profile of scavenged assemblages suggested here, rather, rests on the much more rapid and thorough consumption by initial consumers of young versus adult individuals, and the resulting predominant availability of adults to scavengers. Only young of particularly large species are likely to survive initial consumption frequently and be a component of a scavengers diet. Age data from Plio-Pleistocene archaeological sites at Koobi Fora and Olduvai are now available (Bunn 1982, Potts 1982). No summary data are provided, but most assemblages analyzed by Bunn do seem to be dominated by adults. Whether the proportions of subadult to adult individuals is sufficiently distinct from that expected from age-mortality profiles of live populations is a question that requires further research (cf. Klein 1982).

Characteristic # 2 above, that assemblages accumulated by scavenging should be dominated by larger size classes of herbivores, is consistent with expectations of many researchers. I mentioned in Chapter 1 the reasoning that the mere size of many of the animals represented at archaeological sites seemingly prohibited their hunting, and hence must have been scavenged. The data presented here on size specific carcass persistence and completeness of initial consumption provide the first positive support for this long-held expectation. These data indicate that a scavenger will rarely encounter edible tissues remaining on carcasses of small taxa, and, owing to the greater abundance of medium-sized adults over larger taxa in modern and past ecosystems, will derive the most regular scavenging opportunity from medium-sized adult carcasses. Vrba (1975, 1980) has also suggested a size-based criterion for distinguishing assemblages accumulated by scavenging, noting that, compared to hunted assemblages, scavenged assemblages should contain a larger size range of animals. This suggestion is based on the narrow size range of prey preferred by individual predator taxa, and the expected larger range of carcass sizes available to more eclectic scavengers. Vrba's criterion must be qualified in light of the findings presented here, to account for the typical complete consumption by predators of small taxa, and the availability to scavengers of incompletely consumed carcasses of primarily larger taxa. Bunn's (1982) data from archaeological sites at Koobi Fora and Olduvai show the bone assemblages to be dominated by medium-sized and larger taxa. Given the good preservation of remains at most sites analyzed by Bunn, and the presence of some individuals of smaller taxa, these data are consistent with a primarily scavenged mode of accumulation.

That archaeological faunal assemblages accumulated through scavenging should be dominated by limb and head parts (characteristic # 3 above) requires some discussion. Binford (1984) and Potts (1983) have proposed different sets of body parts to be characteristic of assemblages accumulated through scavenging.

Binford suggests that cranial bones and foot bones (metapodials and phalanges) are the last edible carcass parts to be utilized and should therefore dominate scavenged assemblages. He uses this reasoning to ascribe to scavenging the foot and head bone dominated larger bovid component of Middle Stone Age assemblages from the site of Klasies River Mouth. These observations are similar to mine with one very important exception: in addition to foot bones, upper limb bones (humeri, femora, radius-ulnae, tibiae) should be equally characteristic of scavenged assemblages. Hyenas do usually consume limb bone marrow from the proximal to the distal (foot) end (Table 3.4). Although statistically this might be a basis for expecting a more frequent availability of foot compared to upper limb marrow bones, in actuality, hyenas rarely start chewing a complete limb without consuming marrow from <u>all</u> parts, including foot bones. Regardless, hyenas <u>are</u> very infrequent providers of marrow bones for scavengers. Most marrow bones available are those abandoned by lions, which, on medium-sized and larger carcasses, abandon <u>all</u> limb bones with marrow cavities intact. In light <u>of</u> these observations, I suggest that the foot bone to upper limb bone dominance among the larger bovids at Klasies River Mouth cannot be explained solely by hominid scavenging. If hominids did scavenge most larger bovids at the site, they almost certainly obtained as many upper limb bones as foot bones, such that some additional factor operated to reduce the number of upper limb bones that were originally brought to or preserved at the site (see Blumenschine 1986).

Potts (1983) uses Hill's (1975) carcass disarticulation sequence as a basis for predicting the carcass parts most likely to be available to a scavenger. Specifically, Potts suggests that assemblages accumulated through scavenging should be dominated by parts that disarticulate relatively late in the sequence. These parts include bones from the postcranial axial region of the skeleton relative to limb bones, and among limb parts, hindlimbs relative to forelimbs. Assemblages accumulated through scavenging, according to Potts, should have a high proportion of axial parts relative to limb parts, and a high proportion of hindlimbs relative to forelimbs. In both cases, the latter set of parts are those elements that disarticulate from a carcass earlier in the sequence. There is, however, a fundamental problem with these inferences. Much of the disarticulation sequence occurs well after the resource life of a carcass has been exhausted. After this point, the main disarticulating agents are physical weathering, microorganisms and insects that degrade ligaments that connect bone. These late stages of the disarticulation sequence are irrelevant to what is available to a scavenger seeking nutrients. What is relevant is the sequence by which edible tissues are consumed (Chapter 1, and Blumenschine 1986). According to the general consumption sequence (Chapter 3, Table 3.1) most axial elements save the head are completely utilized (i.e. defleshed) before tissues within bones (limbs and head) are consumed. A high representation of postcranial axial parts in an archaeological bone assemblage is therefore indicative of access by the bone accumulator to flesh parts that are eaten early in the consumption sequence. Since these are usually consumed before abandonment by initial consumers on all but the largest carcasses, such an assemblage is not likely to have been accumulated through scavenging. Likewise, hindlimbs tend to be defleshed and then destroyed before forelimbs. A dominance of forelimb over hindlimb parts in an archaeological accumulation is therefore more likely to be indicative of relatively late access to carcasses (i.e. scavenging), where if any limb parts remain with or without flesh, these are most likely to be forelimbs. The body parts expected to dominate scavenged bone assemblages on the basis of the consumption sequence (limbs, and especially forelimbs, and the head) are, in fact, the converse of those expected on the basis of the disarticulation sequence (axial bones versus few limbs, many hindlimbs relative to forelimbs).

Bunn (1982) notes a very high representation of head and limb parts over postcranial axial parts in most analyzed assemblages from Koobi Fora and Olduvai. Bunn ascribes this pattern to the relative durability and identifiability of the parts. It is interesting to note that it is precisely the more durable carcass parts, i.e. those encased in thick bone, which are most likely to inhibit consumption by many initial consumers and consequently be available to a hungry hominid scavenger wielding a hammerstone. [This simple correlation between differential bone preservability and the accessibility of edible tissues borne by or contained in the bone was recognized long ago by Washburn (1957).] Given this, it can be suggested that the limb and head bone dominance at many early archaeological sites is more consistent with a scavenged than a hunted mode of accumulation.

Data on forelimb to hindlimb ratios for six Olduvai Bed I faunal accumulations presented by Potts (1983) are also more consistent with a scavenged mode of accumulation. Among small and large bovids, forelimb:hindlimb ratios vary in unexplained ways among the sites. However, for the numerically dominant medium-sized taxa, all sites show a clear dominance of forelimbs over hindlimbs. Contrary to Potts' interpretation based on the carcass disarticulation sequence, this pattern is suggestive of scavenging on the carcass size class suggested here to have been most regularly available to hominids. While a scavenged rather than hunted mode of accumulation seems to predominate at these sites from the general body part data cited above, this conclusion is regarded as tentative until the affect of other taphonomic factors unrelated to hominid selection of parts are fully considered (Blumenschine 1986).

The fourth suggested characteristic of bone assemblages accumulated through scavenging relates to the skeletal distribution of defleshing cut marks inflicted by hominids using stone tools. Given the infrequent availability to scavengers of substantial quantities of flesh, one can expect scavenged assemblages to display most defleshing cut marks on bones defleshed last by initial consumers, i.e. lower limb, head and neck bones (Tables 3.1 - 3.4). This is especially the case if scavenging is concentrated on medium-sized adults. In that larger quantities of flesh occasionally remain on carcasses, cut marks should be progressively less common on bones defleshed increasingly earlier in the consumption sequence.

Bunn (1982, 1983) provides some data on the frequency distribution of cut marks on different bones of small and large bovids from the Plio-Pleistocene site of FLK Zinj at Olduvai. Head parts and distal limb bones are among the most commonly cut-marked bones, bearing much higher incidences of cut marks than postcranial axial bones. This is consistent with a scavenged mode of accumulation. However, upper limb bones, which are defleshed relatively early in the consumption sequence by modern carnivores, are also frequently cut-marked at FLK Zinj. This might suggest more common access to flesh than has been suggested to be available to a hominid scavenger. It might, on the other hand, be a result of the removal of small scraps of flesh that are abandoned by most consumers commonly distributed all over the body. Alternatively, the high frequency of cut-marked upper limb bones might not reflect defleshing, but rather disarticulation of bones acquired already defleshed. Currently, these alternatives cannot be tested [notwithstanding Shipman's (1983) oversimplified correlation of cut marks located near joints or on shafts with disarticulation and defleshing, respectively]; experiments are clearly needed to distinguish cut marks inflicted for defleshing versus disarticulating, and for removing whole muscle bundles versus small scraps of flesh. It is interesting to note, however, that among upper limb bones from FLK Zinj, humeri are much more frequently cut-marked than femora of both small and large bovids. This pattern might result from scavenging by hominids, as femora are commonly among the first parts to be defleshed by modern carnivores, with humeri persisting with all flesh until the ribcage and all parts posterior to it (save the tibiae) have been stripped of their meat.

The final suggested characteristic (#5 above) of bone assemblages accumulated through scavenging is a qualification of numbers 3 and 4 above related to size differences among taxa. Here, smaller taxa are expected to be represented more exclusively by those skeletal parts eaten from later in the consumption sequence than larger taxa. Additionally, smaller taxa should bear defleshing cut marks on a more limited series of body parts (i.e. those defleshed last by modern carnivores) than larger taxa. In both cases, the greater likelihood of a scavenger encountering larger carcasses in a more complete resource state than smaller ones forms the basis for these predictions. Data on the relative frequency and placement of cut marks cannot be applied currently for the reasons cited above. However, available skeletal part data are amenable to the above test, as conducted below on data provided by Bunn (1982) for bovids at FLK Zinj.

Table 5.9 shows the number of selected skeletal parts for size 1 and 2 bovids, and size 3 and 4 bovids at FLK Zinj. The body parts used are grouped into two categories. The first category contains those parts typically available with edible tissue only early in the consumption sequence. These parts, including the pelvis, vertebrae, ribs, and scapula, provide only flesh and no within-bone edible tissues, and are expected to have been more commonly available on large carcasses. The second category includes bones that are eaten from only late in the consumption sequence, and, as such, are necessarily restricted to matapodials and phalanges. Upper limb and head parts are excluded from the analysis as these bear flesh and contain edible tissues that are eaten at different stages in the consumption sequence. For all parts used, the total values of lefts and rights provided by Bunn are used. For vertebrae, ribs and metapodials, only the most common portion of these parts given by Bunn are tabulated to minimize artificial and possibly disproportionate inflation of per-part numbers resulting from fragmentation.

The data in Table 5.9 show a marked, size-based asymmetry in the frequency of the two body part classes. Size 1 and 2 bovids are represented by nearly equal numbers of the two skeletal categories, while larger bovids are much more frequently represented by parts available as food sources only relatively early in the consumption sequence. In other words, relative to larger bovids, the small bovid skeletal profile shows a greater degree of skewing toward those parts most frequently available to a scavenger on any carcass size class. The highly significant asymmetry of the data are consistent with suggested characteristic #5 of scavenged bone assemblages, where scavengers are more likely to encounter smaller carcasses with a very limited series of bones remaining with edible tissues.

The application of the five suggested characteristics of scavenged assemblages to some Plio-Pleistocene archaeological bone accumulations are mainly consistent with a scavenged mode of animal food acquisition. The results, however, must be viewed as tentative for two main reasons. First, the criteria were applied in only a general manner due to the current availability of data imperfectly suited to the tests. For instance, the results obtained in Table 5.9 may result not from scavenging, but from size-dependent differential fragmentation; this alternative (deemed less likely than the scavenging interpretation) cannot be tested with currently available data. Second, the above discussion makes no explicit attempt to evaluate other processes of accumulation (e.g. hunting) or disturbance that may have operated on the assemblages. Both sources of inconclusiveness need to be more fully

Table 5.9: Size-specific frequency of bovid skeletal parts at FLK Zinj. The selected series of body parts are those available with edible tissues only early in the consumption sequence, and those eaten from only very late in the consumption sequence. Data are from Bunn (1982, Table 4.28). The difference in the size distribution of grouped parts is highly significant ($X^2 = 28.1$, d.f. = 1, $p < .001$).

	NUMBER OF ELEMENTS	
	Size 1 + 2	Size 3 + 4
Parts eaten from only early		
Pelvis	10	18
Vertebrae (centra only)	15	32
Ribs (proximal only)	17	35
scapula	6	14
Sub-Total	48 (48%)	99 (77.3%)
Parts eaten from only late		
Metapodials (proximal only)	25	17
Phalanges (complete and epiphyseal remains)	27	12
Sub-Total	52 (52%)	29 (22.7%)
TOTAL	100	128

considered (e.g. Blumenschine 1986). Still, the consistency of the results with predictions is suggestive of a possibly important role for scavnging in early hominid subsistence.

Conclusions

The nature of scavenging opportunities suggested for early hominids has some general implications for our understanding of early hominid adaptation, and the meaning of the early archaeological record. If it can be conclusively demonstrated that scavenging by early hominids generated most components of archaeological bone assemblages, several suggestions would apply. First, it can be suggested that marrow bones and the brain made a much more significant dietary contribution relative to flesh than is commonly assumed. Most paleoanthropologists assume flesh to have been the important new source of nutrients available with the advent of animal food consumption by hominids. Marrow and the brain have generally been overlooked as an important and regularly available source of nutrients. Indeed, Shipman (1984) and Potts and Shipman (1981), noting the high frequency of stone tool cut marks on meatless foot bones at early archaeological sites, propose the exploitation of these carcass parts for non-dietary uses, such as the need for tendon in tool manufacture. While this might be true to an extent, it is clear from unpublished experimental studies I have carried out that to extract marrow within foot bones, they must first be skinned, the tendon removed and usually the periosteum scraped away. All of these activities geared to extracting marrow produce cut marks on the bone. Only Binford (1981, 1983) has previously recognized the potentially large role of marrow and the brain in hominid dietary strategies. It is the marrow and head contents, in fact, for which there appears to be an open scavenging niche in certain ecological contexts today. If hominids scavenged exclusively from abandoned felid kills in riparian woodland, flesh would indeed have been relatively unimportant.

If scavenging was regularly practiced in the accumulation of archaeological bone assemblages, it can be suggested that the novel set of behaviors resulting in dense concentrations of stone artifacts and bone were distinctly dry season activities. Speth and Davis (1976) have previously suggested many of the early sites at Olduvai to be dry season sites owing to the presence of large numbers of carapaces of turtles available only during the dry season. The findings on scavenging suggest that all good feeding opportunities, whether from felid kills or natural deaths, consistently occur during the dry season. It therefore becomes possible that archaeologists studying the Plio-Pleistocene record are being informed only of dry season activities, and have little or no information on activities conducted during rainy periods.

Finally, if Plio-Pleistocene hominids can be demonstrated to have scavenged, it is possible that central-place foraging and the suite of other human traits associated with the use of a home base had yet to evolve. This has been suggested to be the case if hominids scavenged from the most readily accessible source, i.e. abandoned felid kills in riparian woodland. The remains of felid kills of medium-sized adults, though providing substantial nourishment for a single individual, is suggested to come in insufficiently large packets both to finance central-place foraging and to afford the surplus so essential to reciprocity systems of the food sharing/home base model of early hominid lifestyles. Archaeological evidence that would be consistent with scavenging exclusively within riparian woodland include the four criteria presented earlier, with emphasis placed on a strong predominance of cut marks on bones defleshed very late by modern felids. Although the current evidence shows cut marks to have a wider skeletal distribution, it is uncertain which of these were inflicted to disarticulate defleshed bone, to remove small scraps of flesh, or to strip away whole muscle bundles.

Scavenging, however, need not always imply routed foraging. During dry periods of high natural mortality, flesh can be scavenged in quantities many times larger than individual daily appetites. The added energetic costs of central-place foraging could not only be financed by the bonanza provided by natural deaths, but the carcasses would also provide a large surplus of food that could be brought to the home base and distributed among group members. Archaeological evidence consistent with the regular scavenging of natural deaths would include a larger range of body parts than those acquired by scavenging abandoned felid kills alone, particularly a high proportion of axial parts in addition to limb parts. The greater diversity of parts should also display a more even skeletal distribution of cut marks inflicted while removing muscle bundles. Finally, limb bones should be recovered largely complete, as marrow of natural deaths is of poor quality and an unattractive source of nutrients both to hominids and scavengers of abandoned hominid sites.

Scavenging natural deaths through central-place foraging does not require the existence of a sexual division of subsistence labor and the associated daily fissioning of a group into foraging sub-groups that reassemble to exchange foods collected during the day. Scavenging natural deaths in open vegetation habitats, in fact, was suggested to have required organized anti-predator tactics, the effectiveness of which would be enhanced by large group sizes. Risk of exposure to predation would still place a premium on return to a riparian refuge, but here, the hominid group would move as a whole unit throughout the daily round. It may be that only the advent of regular hunting, often requiring discreet stalking tactics, that selected for the superimposition upon central-place foraging of daily fissioning of groups into specialized subgroups and other subsistence-related social adjustments seen in modern hunter-gatherers. Central-place foraging can be viewed in this way as a preadaptation to successful hunting, with origins in scavenging.

BIBLIOGRAPHY

Anderson, D.R., and R.S. Pospahala
1970 Correction of bias in belt transect studies of immotile objects. J. Wildl. Manag., 34, 141-146.

Ardrey, R.
1976 The Hunting Hypothesis. Atheneum:New York.

Bartholomew, G.A., and J.B. Birdsell
1953 Ecology and the protohominids. Amer. Anthrop., 55, 481-496.

Behrensmeyer, A.K.
1975 The taphonomy and paleoecology of Plio-Pleistocene vertebrate assemblages east of Lake Rudolf, Kenya. Bull. Mus. Comp. Zool., 146, 473-578.

1983 Patterns of natural bone distribution on recent land surfaces: implications for archaeological site formation. In: J. Clutton-Brock and C. Grigson, eds., 1983, pp. 93-106.

Behrensmeyer, A.K., and D.E. Dechant-Boaz
1980 The recent bones of Amboseli National Park, Kenya, in relation to East African paleoecology. In: A.K. Behrensmeyer and A.P. Hill, eds., 1980, pp. 72-92.

Behrensmeyer, A.K., and A.P. Hill, eds.
1980 Fossils In The Making: Vertebrate Taphonomy and Paleoecology. Univ. Chicago Press:Chicago.

Bell, R.H.V.
1971 A grazing succession in the Serengeti. Scient. Amer., 225, 86-93.

Bertram, B.C.R.
1979 Serengeti predators and their social systems. In: A.R.E. Sinclair and M. Norton-Griffiths, eds., 1979, pp. 221-248.

Bewes, J.W.
1931 The ecological viewpoint. S. African J. Sci., 28, 1-15.

Binford, L.R.
1977 Olorgesaille deserves more than the usual book review. J. Anthrop. Res., 33, 493-502.

1978 Nunamiut Ethnoarchaeology. Academic Press:New York.

1981 Bones: Ancient Men and Modern Myths. Academic Press:New York.

1983 In Pursuit of the Past. Thames and Hudson:London.

1984 Faunal Remains from Klasies River Mouth. Academic Press:New York.

Binford, L.R., and J.B. Bertram
1977 Bone-frequencies and attritional processes. In L.R. Binford, ed., _For Theory Building In Archaeology_. Academic Press:New York, pp. 77-153.

Bonnefille, R.
1976 Palynological evidence for an important change in vegetation of the Omo Basin between 2.5 and 2 million years ago. In: Y. Coppens et al., 1976, pp. 421-431.

1984 Cenozoic vegetation and environments of early hominids in East Africa. In R.O. Whyte, ed., _The Evolution of the East Asian Environment, Vol. II: Palaeobotany, Palaeozoology and Palaeoanthropology_. Univ. Hong Kong Centre of Asian Studies, pp. 579-612.

Blumenschine, R.J.
1986 Carcass consumption sequences and the archaeological recognition of scavenging. (Abstract). Paper presented at a symposium entitled: "The Longest Record: The Human Career in Africa", Berkeley, April 1986.

Blumenschine, R.J., and T.M. Caro
i.p. Unit flesh weights of some East African bovids. _Afr. J. Ecol._

Brain, C.K.
1967 Hottentot food remains and their bearing on the interpretation of fossil bone assemblages. _Scient. Pap. Namib Desert Res. Stn._, no. 32, pp. 1-7.

1969 The contribution of Namib Desert Hottentots to an understanding of australopithecine bone accumulations. _Scient. Pap. Namib Desert Res. Stn._, no. 39, pp. 13-22.

1981 The Hunters Or The Hunted? An Introduction to African Cave Taphonomy. Univ. Chicago Press:Chicago.

Brooks, P.M.
1978 Relationship between body condition and age, growth, reproduction and social status in impala, and its application to management. S. African J. Wildl. Res., 8, 151-157.

Brooks, P.M., J. Hanks, and J.V. Ludbrook
1977 Bone marrow as an index of condition in African ungulates. S. African J. Wildl. Res., 7, 61-66.

Brown, F.H., and W.P. Nash
1976 Radiometric dating and tuff mineralogy of Omo Group deposits. In: Y. Coppens, et al., eds., 1976, pp. 50-63.

Bunn, H.T.
1981 Archaeological evidence for meat-eating by Plio-Pleistocene hominids from Koobi Fora and Olduvai Gorge. Nature, 291, 574-577.

1982 Meat Eating and Human Evolution: Studies on the Diet and Subsistence Patterns of Plio-Pleistocene Hominids in East Africa. Ph.D. Dissertation, University of California, Berkeley.

1983 Evidence on the diet and subsistence patterns of Plio-Pleistocene hominids at Koobi Fora, Kenya, and Olduvai Gorge, Tanzania. In: J. Clutton-Brock and C. Grigson, eds., 1983, pp. 121-130.

Bunn, H.T., J.W.K. Harris, G. Isaac, Z. Kaufulu, E. Kroll, K. Schick, N. Toth, and A.K. Behrensmeyer
1980 FxJj50: an early Pleistocene site in northern Kenya. World Archaeol., 12, 109-136.

Burnham, K.P. and D.R. Anderson
1976 Mathematical models for non-parametric inferences from line transect data. Biometrics, 32, 325-336.

Butynski, T.M.
1982 Vertebrate predation by primates: a review of hunting patterns and prey. J. Human Evolution, 11, 421-430.

Campbell, B.
1979 Ecological factors and social organization in human evolution. In: I.S. Bernstein and E.O. Smith, eds., Primate Ecology and Human Origins. Garland STPM Press:New York, pp. 291-312.

Carr, C.J.
1976 Plant ecological variation and pattern in the Lower Omo Basin. In: Y. Coppens, et al., eds., 1976, pp. 432-467.

Churcher, C.S.
1982 Oldest ass recovered from Olduvai Gorge, Tanzania, and the origin of asses. J. Paleontol., 56, 1124-1132.

Clark, J.D.
1959 The Prehistory of Southern Africa. Penguin:Baltimore.

1960 Human ecology during the Pleistocene and later times in Africa south of the Sahara. Curr. Anthrop., 1, 307-324.

Clutton-Brock, J., and C. Grigson, eds.
1983 Animals and Archaeology: 1. Hunters and Their Prey. British Archaeological Reports International Series, no. 163, Oxford.

Coe, M.
1980 The role of modern ecological studies in the reconstruction of paleoenvironments in sub-Saharan Africa. In: A.K. Behrensmeyer and A.P. Hill, eds., 1980, pp. 55-67.

Coe, M.J., D.H. Cumming, and J. Phillipson
1976 Biomass and production of large African herbivores in relation to rainfall and primary production. Oecologia, 22, 341-354.

Coppens, Y., and F.C. Howell
1976 Mammalian faunas of the Omo Group: distributional and biostratigraphic aspects. In: Y. Coppens et al., eds., 1976, pp. 177-192.

Coppens, Y., F.C. Howell, G.L. Isaac, and R.E. Leakey, eds.
1976 Earliest Man and Environments in the Lake Rudolf Basin: Stratigraphy, Paleoecology and Evolution. Univ. Chicago Press:Chicago.

Corfield, T.F.
1973 Elephant mortality in Tsavo National Park, Kenya. E. African Wildl. J., 11, 339-368.

Dart, R.A.
1953 The predatory transition from ape to man. Intntl. Anthrop. Ling. Rev., 1, 201-219.

deCalesta, D.S., J.G. Nagy, and J.A. Bailey
1977 Experiments on starvation and recovery of mule deer does. J. Wildl. Manag., 41, 81-86.

DeVore, I., and S.L. Washburn
1963 Baboon ecology and human evolution. In F.C. Howell and F. Bourliere, eds., African Ecology and Human Evolution. Viking Fund Publ. in Anthropology, no. 36, pp. 335-367.

Dorst, J., and M.P. Dandelot
1969 A Field Guide to the Larger Mammals of Africa. Houghton Mifflin:Boston.

Dunham, K.M., and M.G. Murray
1982 The fat reserves of impala, *Aepyceros melampus*. *African J. Ecology*, 20, 81-87.

Eaton, R.L.
1974 The Cheetah: The Biology, Ecology and Behavior of an Endangered Species. van Nostrand Reinhold:New York.

Eaton, R.L.
1979 Interference competition among carnivores: model for the evolution of social behavior. *Carnivore*, 2, 9-16.

Eberhardt, L.L.
1978 Transect methods for population studies. *J. Wildl. Manag.*, 42, 1-31.

Elliott, J.P., I. McTaggart-Cowan, and C.S. Holling
1977 Prey capture by African lion. *Canadian J. Zool.*, 55, 1811-1828.

Elliott, J.P., and I. McTaggart-Cowan
1978 Territoriality, density and prey of lion in Ngorongoro Crater, Tanzania. *Canadian J. Zool.*, 56, 1726-1734.

Estes, R.D., and R.K. Estes
1979 Birth and survival of wildebeest calves. *Z. Tierpscyh.*, 50, 45-95.

Ewer, R.F.
1967 The fossil hyaenids of Africa - a reappraisal. In: W.W. Bishop and J.D. Clark, eds., *Background to Evolution In Africa*. Univ. Chicago Press:Chicago. pp. 109-123.

Frame, G.W.
1986 Carnivore Competition and Resource Use in the Serengeti Ecosystem of Tanzania. Ph.D. Dissertation, Utah State Univ., Logan.

Franzman, A.W., and P.D. Arneson
1976 Marrow fat in Alaskan moose femurs in relation to mortality factors. *J. Wildl. Manag.*, 40, 336-339.

Gentry, A.W.
1976 Bovidae of the Omo Group deposits. In: Y. Coppens et al., eds., 1976, pp. 275-292.

Gentry, A.W., and A. Gentry
1978 Fossil bovidae (mammalia) of Olduvai Gorge, Tanzania, Part I. Bull. Brit. Mus. Nat. Hist., London. vol. 29.

Goodall, J. and D.A. Hamburg
1975 Chimpanzee behavior as a model for the behavior of early man: new evidence of possible origins of human behavior. Amer. Hdbk. Psychiatry, 6, 14-43.

Hamilton, W.J. III, and K.E.F. Watt
1970 Refuging. Ann. Rev. Ecol. Syst., 1, 263-286.

Hanby, J.P., and J.D. Bygott
1979 Population changes in lions and other predators. In: A.R.E. Sinclair and M. Norton-Griffiths, eds., 1979, pp. 249-262.

Hanks, J., D.H.M. Cumming, J.L. Orpen, D.F. Parry, and H.B. Warren
1976 Growth, condition and reproduction in the impala ram (Aepyceros melampus). J. Zool., 179, 421-435.

Harding, R.S.O.
1973 Predation by a troop of olive baboons (Papio anubis). Amer. J. Phys. Anthrop., 38, 587-591.

Harding, R.S.O., and S.C. Strum
1976 Predatory baboons of Keyopey. Nat. Hist., 85(3):46-53.

Harris, D.
1945 Symptoms of malnutrition in deer. J. Wildl. Manag., 9, 319-322.

Harris, J.M.
1976 Bovidae from the East Rudolf succession. In Y. Coppens et al., eds., 1976, pp. 293-301.

1978 Paleontology. In: M.G. Leakey and R.E. Leakey, eds., Koobi Fora Research Project, Volume 1: The Fossil Hominids and an Introduction to Their Context, 1968-1974. Clarendon Press:Oxford. pp. 32-63.

Harris, J.M., and T.D. White
1979 Evolution of the Plio-Pleistocene African suidae. Trans. Amer. Philos. Soc., vol. 69 no. 2.

Hasegawa, T., M. Hiraiwa, T. Nishida, and H. Takasaki
1983 New evidence on scavenging behavior in wild chimpanzees. Curr. Anthrop., 24, 231-232.

Hausfater, G.
1976 Predatory behavior of yellow baboons. Behavior, 56, 44-68.

Hay, R.L.
1976 Geology of the Olduvai Gorge. Univ. California Press:Berkeley.

Hendey, Q.B.
1974 The Late Cenozoic carnivora of the South-Western Cape Province. Ann. S. African Mus., 63, 1-369.

Hill, A.P.
1975 Taphonomy of Contemporary and Late Cenozoic East African Vertebrates. Ph.D. Dissertation, Univ. London.

1979 Disarticulation and scattering of mammal skeletons. Paleobiol., 5, 261-274.

Hill, A., and A.K. Behrensmeyer
1985 Natural disarticulation and bison butchery. Amer. Antiquity, 50, 141-145.

Hill, K.
1982 Hunting and human evolution. J. Human Evol., 11, 521-544.

Hillman, J.C., and A.K.K. Hillman
1977 Mortality of wildlife in Nairobi National Park during the drought of 1973-1974. E. African Wildl. J., 15, 1-18.

Hooijer, D.A.
1969 Pleistocene East African rhinoceroses. In: L.S.B. Leakey, ed., Fossil Vertebrates of Africa, Volume 1. Academic Press:New York. pp. 71-98.

Houston, D.C.
1972 The Ecology of Serengeti Vultures. Ph.D. Dissertation, Oxford Univ. (Trinity College).

1974 The role of Griffon vultures Gyps africanus and Gyps ruppelli as scavengers. J. Zool., 172, 35-46.

1979 The adaptations of scavengers. In A.R.E. Sinclair and M. Norton-Griffiths, eds., 1979, pp. 263-286.

Howell, F.C.
1968 Early Man. Time-Life Books.

Howell, F.C., and G. Petter
1976 Carnivora from Omo Group Formation, southern Ethiopia. In: Y. Coppens et al., eds., 1979, pp. 314-331.

Isaac, G.Ll.
1971 The diet of early man: aspects of archaeological evidence from Lower and Middle Pleistocene sites in Africa. _World Archaeol._, 2, 278-299.

1976 The activities of early African hominids. In: G.Ll. Isaac and E.R. McCown, eds., _Human Origins: Louis Leakey and the East African Evidence_, W.A. Benjamin:Menlo Park. pp. 483-514.

1978a Food sharing and human evolution: archaeological evidence from the Plio-Pleistocene of East Africa. _J. Anthrop. Res._, 34, 311-325.

1978b The food sharing behavior of protohuman hominids. _Scient. Amer._, 238, 90-108.

1980 Casting the net wide: a critical review of archaeological evidence for early hominid land-use and ecological relations. In: L-K Konigsson, ed., _Current Argument on Early Man_, Pergamon Press:Oxford. pp. 226-253.

1981 Emergence of human behaviour patterns: archaeological tests of alternative models of early hominid behaviour - excavation and experiments. _Philos. Trans. Roy. Soc. Lond._, B292, 177-188.

1983 Bones in contention: competing explanations for the juxtaposition of Early Pleistocene artifacts and faunal remains. In: J. Clutton-Brock and C. Grigson, 1983.

i.p. Koobi Fora Research Project, Volume II: Archaeology. (title approximate).

Isaac, G.Ll., and J.W.K. Harris
1978 Archaeology. In: M.G. Leakey and R.E. Leakey, eds., _Koobi Fora Research Project, Volume 1: The Fossil Hominids and an Introduction to Their Context, 1968-1974_. Clarendon Press:Oxford. pp. 64-85.

Isaac, G.Ll, and D.C. Crader
1981 To what extent were early hominids carnivorous? An archaeological perspective. In: R.S. Harding and G. Teleki, eds., _Ominovorous Primates: Gathering and Hunting in Human Evolution_. Columbia Univ. Press:New York. pp. 37-103.

Isaac, G.Ll., J.W.K. Harris, and F. Marshall
1981 Small is informative: the application of the study of mini-sites and least effort criteria in the interpretation of the Early Pleistocene archaeological record at Koobi Fora, Kenya. Paper circulated at Union Internacional de Ciencias Prehistoricas y Protohistoricas; X Congress. Mexico City, October 1981.

Jarman, P.J., and A.R.E. Sinclair
1979 Feeding strategy and the pattern of resource-partitioning in ungulates. In: A.R.E. Sinclair and M. Norton-Griffiths, eds., 1979, pp. 130-163.

Jonas, D.F.
1976 On an `alternative paleobiology' and the concept of a scavenging phase. Curr. Anthrop., 17, 144-145.

Kappelman, J.
1984 Plio-Pleistocene environments of Bed I and lower Bed II, Olduvai Gorge, Tanzania. Palaeogeogr., Palaeoclimat., Palaeoecol., 48, 171-196.

King, G.E.
1975 Socioterritorial units among carnivores and early hominids. J. Anthrop. Res., 31, 69-87.

1976 Socioterritorial units and interspecific competition: modern carnivores and early hominids. J. Anthrop. Res., 32, 276-284.

Klein, R.G.
1982 Age (mortality) profiles as a means of distinguishing hunted species from scavenged ones in Stone Age archaeological sites. Paleobiol., 8, 151-158.

Kortlandt, A.
1980 How might early hominids have defended themselves against large predators and food competitors? J. Human Evol., 9, 97-112.

Kruuk, H.
1967 Competition for food between vultures in East Africa. Ardea, 55, 171-193.

1972 The Spotted Hyena: A Study of Predation and Social Behavior. Univ. Chicago Press:Chicago.

Kurland, J.A., and S.J. Beckerman
1985 Optimal foraging and hominid evolution: labor and reciprocity. Amer. Anthrop., 87, 73-93.

Kurten, B.
1968 Pleistocene Mammals of Europe. Aldine:Chicago.

Lampretch, J.
1981 The function of social hunting in larger terrestrial carnivores. Mammal. Rev., 11, 169-179.

Lancaster, J.B.
1975 Primate Behavior and the Emergence of Human Culture. Holt, Rhinehart and Winston:New York.

Leakey, L.S.B.
1967 Development of aggression as a factor in early human and prehuman evolution. In: C. Clements and D. Lindsley, eds., Aggression and Defense. Univ. California Press:Los Angeles. pp. 1-33.

Leakey, M.D.
1971 Olduvai Gorge: Excavations in Beds I and II, 1960-1963. Cambridge Univ. Press:Cambridge.

Leakey, M.G.
1976 Carnivora of the East Rudolf succession. In Y. Coppens et al., eds., 1976. pp. 302-313.

Ledger, H.P.
1968 Body composition as a basis for a comparative study of some East African mammals. Symp. Zool. Soc. Lond., 21, 289-310.

Lee, R.B.
1968 What hunters do for a living, or how to make out on scarce resources. In: R.B. Lee and I. DeVore, eds., Man the Hunter. Aldine:Chicago. pp. 30-48.

Maddock, L.
1979 The 'migration' and grazing succession. In: A.R.E. Sinclair and M. Norton-Griffiths, eds., 1979, pp. 104-129.

Maglio, V.J.
1978 Patterns of faunal evolution. In: V.J. Maglio and H.B.S. Cooke, eds., Evolution of African Mammals. Harvard Univ. Press:Cambridge. pp. 603-619.

Maguire, J.M., D. Pemberton, and M.H. Collett
1980 The Makapansgat Limeworks grey breccia: hominids, hyaenas, hystricids or hillwash? Paleontol. Africana, 23, 75-98.

McNaughton, S.J.
1976 Serengeti migratory wildebeest: facilitation of energy flow by grazing. Science, 191, 92-94.

1979 Grassland-herbivore dynamics. In: A.R.E. Sinclair and M. Norton-Griffiths, eds., 1979, pp. 46-81.

Maynard-Smith, J.
1974 Models in Ecology. Cambridge Univ. Press:Cambridge.

Merrick, H.V., and J.P.S. Merrick
1976 Archaeological occurrences of earlier Pleistocene age from the Shungura Formation. In: Y. Coppens et al., eds., 1976, pp. 574-584.

Milton, K.
1981 Distribution patterns of tropical plant foods as an evolutionary stimulus to primate mental development. _Amer. Anthrop._, 83, 534-548.

Morris, K., and J. Goodall
1977 Competition for meat between chimpanzees and baboons of the Gombe National Park. _Fol. Primatol._, 28, 109-121.

Orians, G.H., and N.E. Pearson
1979 On the theory of central place foraging. In: D.J. Horn and G.R. Stairs, eds., _Analysis of Ecological Systems_, Ohio State Univ. Press:Colombus. pp. 155-177.

Pennycuick, L.
1975 Movements of migratory wildebeest population in the Serengeti area between 1960 and 1973. _E. African Wildl. J._, 13, 65-87.

Petter, G.
1973 Carnivores Pleistocenes du Ravin d'Olduvai (Tanzanie). In: L.S.B. Leakey, R.J.G. Savage, and S.C. Coryndon, eds., _Fossil Vertebrates of Africa, Volume 3_, Academic Press:London. pp. 43-100.

Pfeiffer, J.E.
1969 The Emergence of Man. Harper and Row:New York.

Potts, R.B.
1982 Lower Pleistocene Site Formation and Hominid Activities at Olduvai Gorge, Tanzania. Ph.D. Dissertation, Harvard Univ.

1983 Foraging for faunal resources by early hominids at Olduvai Gorge, Tanzania. In J. Clutton-Brock and C. Grigson, eds., 1983, pp. 51-62.

1984 Home bases and early hominids. _Amer. Scient._, 72, 338-347.

Potts, R.B., and P. Shipman
1981 Cutmarks made by stone tools on bones from Olduvai Gorge, Tanzania. Nature, 291, 577-580.

Ransom, A.B.
1965 Kidney and marrow fat as indicators of white-tailed deer condition. J. Wildl. Manag., 29, 397-398.

Read-Martin, C.E., and D.W. Read
1975 Australopithecine scavenging and human evolution: an approach from faunal analysis. Curr. Anthrop., 16, 359-368.

Richardson, P.R.K.
1980 Carnivore damage to antelope bones and its archaeological implications. Paleontol. Africana, 23, 109-125.

Robinette, W.L., D.A. Jones, J.S. Gashwiler, and C.M. Aldous
1954 Method for censusing winter-lost deer. Trans. N. Amer. Wildl. Conf., 19, 511-525.

Robinette, W.L., C.M. Loveless, and D.A. Jones
1974 Field tests of strip census methods. J. Wildl. Manag., 38, 81-96.

Rodman, P.S., and H.M. McHenry
1980 Bioenergetics and the origin of hominid bipedalism. Amer. J. Phys. Anthrop., 52, 103-106.

Schaller, G.B.
1972 The Serengeti Lion: A Study of Predator-Prey Relations. Univ. Chicago Press:Chicago.

Schaller, G.B., and G.R. Lowther
1969 The relevance of carnivore behavior to the study of early hominids. S. W. J. Anthrop., 25, 307-341.

Sept, J.M.
1984 Plants and Early Hominids in East Africa: A Study of Vegetation in Situations Comparable to Early Archaeological Site Locations. Ph.D. Dissertation, University of California, Berkeley.

Shipman, P.
1983 Early hominid lifestyle: hunting and gathering or foraging and scavenging? In: J. Clutton-Brock and C. Grigson, eds., 1983, pp. 31-49.

1984 Scavenger hunt. Nat. Hist., 93(4):20-27.

Shipman, P., and J.E. Phillips
1976 On scavenging by hominids and other carnivores. *Curr. Anthrop.*, 17, 170-172.

Sinclair, A.R.E.
1975 The resource limitation of trophic levels in tropical grassland ecosystems. *J. Anim. Ecol.*, 44, 497-520.

1979a The Serengeti environment. In: A.R.E. Sinclair and M. Norton-Griffiths, eds., 1979, pp. 31-45.

1979b Dynamics of the Serengeti ecosystem. In: A.R.E. Sinclair and M. Norton-Griffiths, eds., 1979, pp. 1-30.

Sinclair, A.R.E., and P. Duncan
1972 Indices of condition in tropical ruminants. *E. African Wildl. J.*, 10, 143-149.

Sinclair, A.R.E., and M. Norton-Griffiths, eds.
1979 Serengeti: Dynamics of an Ecosystem. Univ. Chicago Press:Chicago.

Smith, N.S.
1970 Appraisal of condition estimation methods for East African ungulates. *E. African Wildl. J.*, 8, 123-129.

Speth, J.D.
1983 Bison Kills and Bone Counts: Decision-Making by Ancient Hunters. Univ. Chicago Press:Chicago.

i.p. Early hominid subsistence strategies in seasonal habitats. Submitted to *J. Archaeol. Science*.

Speth, J.D., and D.D. Davis
1976 Seasonal variability in early hominid predation. *Science*, 192, 441-445.

Speth, J.D., and K.A. Speilman
1983 Energy source, protein metabolism and hunter-gatherer subsistence strategies. *J. Anthrop. Archaeol.*, 2, 1-31.

Strum, S.C.
1975 Primate predation: interim report on the development of a tradition in a troop of olive baboons. *Science*, 187, 755-757.

1983 Baboon cues for meat eating. *J. Human Evol.*, 12, 327-336.

Szalay, F.S.
1975 Hunting-scavenging protohominids: a model for human origins. *Man*, 10, 420-429.

Teleki, G.
1973 The omnivorous chimpanzee. *Scient. Amer.*, 228, 32-42.

1975 Primate subsistence patterns: collector-predators and gatherer-hunters. *J. Human Evol.*, 4, 125-184.

1981 The omnivorous diet and eclectic feeding habits of chimpanzees in Gombe National Park, Tanzania. In: R.S.O. Harding and G. Teleki, eds., *Omnivorous Primates*. Columbia Univ. Press:New York. pp. 303-343.

Thompsom, P.R.
1975 Cross species analysis of carnivore, primate and hominid behavior. *J. Human Evol.*, 42, 113-124.

1976 Behavioral model for *Australopithecus africanus*. *J. Human Evol.*, 5, 547-558.

van Couvering, J.A.H.
1980 Community evolution in East Africa during the Late Cenozoic. In A.K. Behrensmeyer and A.P. Hill, eds., 1980, pp. 272-298.

van Lawick-Goodall, J.
1968 The behavior of free-living chimpanzees in the Gombe Stream Reserve. *Anim. Behav. Monogr.*, 1, 161-311.

Vincent, A.S.
1985 Wild Tubers as a Harvestable Resource in the East African Savannas: Ecological and Ethnographic Studies. Ph.D. Dissertation, University of California, Berkeley.

Voorhies, M.
1969 Taphonomy and population dynamics of an Early Pliocene vertebrate fauna, Knox County, Nebraska. *Contrib. Geol.*, special paper no. 1, Univ. Wyoming Press.

Vrba, E.S.
1975 Some evidence of chronology and paleoecology of Sterkfontein, Swartkrans and Kromdraai from the fossil bovidae. *Nature*, 254, 301-304.

1980 The significance of bovid remains as indicators of environment and predation patterns. In: A.K. Behrensmeyer and A.P. Hill, eds., 1980, pp. 247-271.

Washburn, S.L.
1957 Australopithecines: the hunters or the hunted? *Amer. Anthrop.*, 59, 612-614.

Washburn, S.L., and C.S. Lancaster
1968 The evolution of hunting. In: R.B. Lee and I. DeVore, eds., Man the Hunter. Aldine:Chicago. pp. 293-303.

Western, D.
1975 Water availability and its influence on the structure and dynamics of a savanna large mammal community. E. African Wildl. J., 13, 265-286.

Wind, J.
1976 On scavenging in an alternative paleobiological concept. Curr. Anthrop., 17, 774-775.

Woodburn, J., (discussant)
1968 Discussion, Part VIIb: primate behavior and the evolution of aggression. In: R.B. Lee and I. DeVore, eds., Man the Hunter. Aldine:Chicago. pp. 339-344.

Wrangham, R.W.
1977 Feeding behavior of chimpanzees in Gombe National Park, Tanzania. In: T.H. Clutton-Brock, ed., Primate Ecology: Feeding and Ranging Behavior of Lemurs, Monkeys and Apes. Academic Press:New York. pp. 504-538.

www.ingramcontent.com/pod-product-compliance
Lightning Source LLC
LaVergne TN
LVHW071213210926
848482LV00015B/796